Cost and Project Management of Modern Methods of Construction

This textbook is aimed at providing students with an insight into the complexities of project management and cost management when working with Modern Methods of Construction (MMC) such as off-site and volumetric construction.

The book begins with an introduction to the MMC sector and educates readers on what is meant by modern methods of construction. Chapters then cover procurement, supply chain management, quality management, and time management. The book then delves into the detail of cost management by introducing the basics of construction economics, cost variable analysis and the overall impact of introducing MMC into cost management.

Through a series of easy-to-follow discussions, case studies from real-life industry experience, the authors clearly communicate the complexities of finance, contract, and project management when using modern methods of construction.

Cost and Project Management of Modern Methods of Construction is an essential new textbook for undergraduate and further education students on construction, architecture, built environment, and quantity surveying courses, but is also a handy guide for practicing quality surveyors and project managers dealing with an increasingly MMC focused industry.

David Roberts has several years' experience working in the modern construction methods sector and in higher education. He worked in industry for over 15 years, eventually owning and running two business that were related to modern methods of construction, a timber frame manufacturing and installation company and a modular company start up. Since working in the higher education sector, he has worked delivering quantity surveying, cost management and project management degrees and higher degrees. Recent areas of research have focused on modern methods of construction, particularly modular production and its financial impact and costs surrounding the industry.

Gopalakrishnan Gomathinayagam is BIM Manager at L&Q and was course director for the BSc Construction Management, lecturer in Construction at Coventry University, and online tutor for the MSc Project Management for Construction at the University of Portsmouth. He completed his Bachelor's in Civil Engineering, before gaining several years of international site engineering and project management experience. As a passionate learner, he then entered the world of academia to share his knowledge and experience. He specialised throughout his post graduate learning in Building Information Modelling and digital construction. He is a Fellow of the Higher Education Academy and his research and teaching focusses on the application of artificial intelligence and BIM to enhance construction performance at various stages of the construction process.

Cost and Project Management of Modern Methods of Construction

David Roberts and Gopalakrishnan Gomathinayagam

Routledge
Taylor & Francis Group
LONDON AND NEW YORK

Designed cover image: Shutterstock.com

First published 2026
by Routledge
4 Park Square, Milton Park, Abingdon, Oxon OX14 4RN

and by Routledge
605 Third Avenue, New York, NY 10158

Routledge is an imprint of the Taylor & Francis Group, an informa business

British Library Cataloguing-in-Publication Data
A catalogue record for this book is available from the British Library

ISBN: 9781032943817 (hbk)
ISBN: 9781032916064 (pbk)
ISBN: 9781003570431 (ebk)

DOI: 10.1201/9781003570431

Typeset in Times New Roman
by KnowledgeWorks Global Ltd.

Contents

Preface

The construction industry is undergoing a profound transformation. Pressures on cost, time, sustainability, and quality have challenged traditional practices and accelerated the adoption of Modern Methods of Construction (MMC). This book was written to provide clarity and practical insight into this evolving landscape.

Our motivation for writing this text stems from the fact we have both worked in academia and industry for a number of years and have noticed a gap between theory and practice within the industry when considering the use of Modern Methods of Construction in projects. This gap has developed between the promise of innovation and the realities of project delivery. While MMC is often presented as a panacea, its successful implementation depends on informed decisions, robust planning, and a clear understanding of its implications for design, procurement, supply chain management, and cost modelling. This book aims to bridge that gap.

The chapters are structured to guide readers through the key considerations:

- Chapter 1 sets the context, exploring why change is necessary and the challenges facing traditional construction.
- Chapter 2 introduces the MMC framework and its categories, providing a foundation for understanding the options available.
- Chapter 3 examines procurement strategies and their critical role in enabling MMC adoption.
- Chapter 4 addresses supply chain management, highlighting the importance of early engagement and digital integration.
- Subsequent chapters focus on *quality, time*, and *cost*, three pillars that underpin project success, before concluding with practical tools for decision-making.

This book is intended for students, practitioners, and decision-makers who seek not only to understand MMC but to apply it effectively. It does not advocate for MMC as a universal solution; rather, it provides the knowledge and frameworks needed to evaluate when and how these methods can deliver value.

Writing this book has been both challenging and rewarding. It reflects countless hours of research, discussion, and collaboration. We hope it serves as a useful resource for those navigating the complexities of modern construction and contributes to the ongoing dialogue about how we build better, faster, and more sustainably.

Dave Roberts and Gopal Gomathinayagam
Coventry University
December 2025

Acknowledgements

From David

I would like to express my sincere gratitude to my co-author, Gopal, for his patience, constructive input, and invaluable insights throughout this project. His collaboration has strengthened both the depth and clarity of this book.

Special thanks also go to Angie and Aven for their unwavering support and understanding during the long hours of research and writing. Your encouragement has been a constant source of motivation.

From Gopal

I am grateful to my co-author, Dave Roberts, for his support, encouragement, and collaboration throughout this work. His insight and support have shaped my contribution to this book and strengthened the work as a whole.

My deepest gratitude goes to my father and my sister, whose steady belief has shaped much of who I am. Above all, my heartfelt thanks to Shomina, Kavin, and Shenba. Your patience, love, and support through every late night and long weekend made this book possible. This journey belongs to you as much as to me.

Dedications

From David

I would like to dedicate this book to Angie and Aven—thank you for your love, patience, and the space you gave me to complete this work. Your support made every page possible.

From Gopal

To my late mother, whose fight against cancer showed me what true strength looks like. Though she is no longer here, her courage and kindness continue to shape every part of my life.

Chapter 1

Introduction

The Construction Industry

The UK construction industry is undergoing a period of significant challenge and change. A well-documented shortage of skilled labour has coincided with growing demands for higher productivity and output. This creates an urgent need to explore new methods of delivering the built environment efficiently and effectively. As one of the largest sectors in the UK economy, contributing around 5.6% to the nation's Gross Domestic Product (GDP) in 2023, the construction sector's performance is closely monitored by government, industry bodies, and the public.

Meeting targets in many areas of construction has proved difficult. For example, the UK Government's ambition to deliver 300,000 new homes per year by the mid-2020s has been hampered by constraints such as limited land availability, complex planning systems, a fragmented supply chain, and persistent skills shortages. By late 2022, it was formally acknowledged that this housing target was unlikely to be met.

Although recent data from 2024 shows a growth in first time buyers there is also a growth in private and social rental sector (MHCLG, 2025). Two hundred and one thousand new builds were completed in the 12 months to June 2025, which is still short of the targeted 300,000. There are often cited reasons for not meeting the housing demand, which can include Lack of available land, planning permission restrictions, lack of available skills to build the homes, etc. It also should be considered from an economic perspective in that if the housing crisis is removed or reduced, then the profit margins for developers will also reduce following supply/demand principles. Therefore, there is a question if the housing system in the UK is set up in a way to perpetuate the issue rather than solve the issue.

One approach gaining increasing attention as a potential solution to these challenges is off-site construction. This involves manufacturing components, or in some cases entire modules, away from the construction site before transporting them for installation. Off-site methods can be categorised into several types:

Panelised systems, where flat, two-dimensional elements are produced and assembled on-site.

Volumetric systems, where complete three-dimensional sections are manufactured in a factory and delivered to site.

Hybrid systems, which combine elements of both approaches and may include pre-fabricated service units or utility pods.

DOI: 10.1201/9781003570431-1

When large portions of a building are produced as complete 3D modules, the process is often referred to as modular construction. This method has attracted growing interest due to its potential to reduce on-site labour requirements, improve quality control, and accelerate project delivery.

A factory-based production environment offers several advantages: improved supervision, consistent quality through standardised processes, reduced exposure to weather conditions, and enhanced safety. However, these benefits are not always reflected in project cost analyses, as they can be difficult to quantify in monetary terms. Cost remains one of the primary factors influencing the adoption of construction methods, and developers are generally more familiar with the financial implications of traditional techniques than with those of modular or other off-site methods. While there is reasonable data on manufacturing and installation costs, less attention has been given to other cost variables, both internal (within a developer's control) and external (outside their influence),that affect overall project viability.

Estimating costs in construction is further complicated by factors such as complex supply chains, varying discount structures during procurement, and the wide range of cost modelling approaches available. Choosing an appropriate cost model early in the design stage is essential to making informed strategic decisions between traditional and off-site approaches. At present, many costing practices are better suited to reporting historical expenditure than to guiding proactive decision-making.

Developments in the construction sector are common and innovative, the research carried out in educational establishments allow for speedy development. Often using engineering and manufacturing principles and trying to adapt them for construction it has allowed for developments such as robotic bricklayers, 3D printing, self-driving excavators, etc. – some of these are seen as quite gimmicky and may never come to fruition, but some of the developments are quickly becoming main stream. One of the issues in the industry is that while research and development is a big part of the sector, it takes a very long time for these developments to be trusted and implemented in standard practice. For example, Computer Aided Design was developed in MIT in the 1950s, but it did not start to become mainstream until the late 1980s, early 1990s when AutoCad entered the market place. Trussed roofs have been around since the 1940s but only became mainstream in the 1980s again. So it takes a while to go from drawing board to practice.

Building Information Modelling (BIM), or sometimes called Building Information Management (in an attempt to move the focus away from a 3D model perception) is a building process where digital information is developed for all of the building components in their entirety, prior to starting on site. This will include a 3D digital model of the building that can be developed with time, cost, quality, energy consumption, clash detection all included in the digital model. If the building is then produced in an exact physical replica of the digital model, then it should (in theory) be perfect and to the client's specification. There are lots of innovative technologies that have been developed on the back of this process.

For example, virtual reality is now used to fully immerse clients in digital buildings prior to construction which will in turn reduce or eliminate the costly and time-consuming client changes. Power tools, excavators and other pieces of equipment can have the digital model processed into them so that they will only operate when in the right place – Hilt, for example, have a nail gun that will only allow the nail to be inserted in the correct location. All of this is also a good advocate for the use to modern methods of construction, off-site construction and/or DfMA. Prefabrication is a process of manufacturing items in a factory setting to a perceived higher level of accuracy than can be achieved on site. Therefore, with designs now needing to be far more accurate so that they emulate the digital design more closely, prefabrication should, in theory, be able to provide this.

This book aims to explore this principle and concept in more detail, discussing how the good and bad aspects of traditional and off-site methods can combine to create unfavourable outcomes, but also if used correctly can enhance the outcomes for clients and all other stakeholder in the process.

Background

Challenges in Traditional Construction

Traditional construction processes, which dominate much of the UK industry, face multiple challenges that can lead to lower productivity, increased costs, and reduced collaboration. A defining feature of conventional methods is the reliance on high levels of manual labour, both skilled and semi-skilled, at the construction site (Farmer, 2016). While essential, on-site labour introduces several human-related risks. These include health and safety hazards such as accidents and work-related illness, productivity fluctuations linked to physical and emotional well-being, and delays caused by external factors such as adverse weather.

Human factors also contribute to variability in quality and cost. For example, errors made during construction can result in rework, wasted materials, and programme delays. In addition to these human risks, traditional site-based production faces logistical challenges. These include the need to coordinate deliveries from numerous suppliers, manage multiple subcontractors, and store materials in outdoor environments, where exposure to weather can affect the durability and performance of building components.

Estimating costs under such conditions is complex. Researchers have noted that cost modelling in construction is hampered by limited availability of reliable data and a reluctance among contractors to share information, often due to competitive sensitivities. This lack of transparency has contributed to the limited understanding of cost variables, particularly in relation to modular and other off-site approaches. Given that cost remains one of the most significant barriers to adopting off-site techniques, improving knowledge in this area is vital.

One of the key challenges facing the move from traditional on-site activities to a more off-site, or modern methods approach, is the Client and the Design team wanting to have a bespoke aesthetic, but quickly, and with good quality, and often these two aspects do not align. As will be seen later in this chapter, cost is often the first consideration when making decisions in a construction project, it is also often the deciding factor. Unfortunately cost and cost certainty are very elusive in construction projects due to the number of elements and cost variables associated with these complex projects. Reducing and standardising the number of elements will make costing clearer, and MMC is a good way of doing this, however it is perceived that the use of MMC will often remove the bespoke nature of clients' aesthetic desires (Goulding and Rahimian, 2020). There is often a trade off when using MMC in projects between maximising the benefits of off-site construction while not reducing the aesthetic appeal that can be achieved.

Costing Practices in Traditional Construction

In the UK, traditional construction projects are typically costed using a Resource Cost Analysis (RCA) approach. This method allocates costs to three main resource categories: labour, plant, and materials. RCA is deeply embedded in industry practice and underpins many established costing systems, such as the Building Cost Information Service (BCIS), which supports pre-tender cost planning, and industry-standard price books used by contractors and developers to inform estimates on future projects (Towey, 2018).

While RCA is well-suited to the multi-stakeholder, site-based nature of traditional construction, it may be less appropriate for certain off-site projects. For example, in modular construction, large portions of the work may be supplied by a single manufacturer, potentially making traditional resource-based costing less precise than alternative methods. Furthermore, a lack of understanding among design teams about how standardisation in modular construction can lead to cost efficiencies means that opportunities for savings are sometimes missed. Even small design adjustments, such as modifying dimensions to align with modular manufacturing standards, can make off-site options more competitive.

Wider Challenges Facing the Construction Sector

The UK construction sector plays a central role in the economy and in meeting society's housing, infrastructure, and welfare needs. However, it faces a number of interrelated challenges:

- *Legislation*: The industry is subject to extensive regulation, covering building safety, planning, environmental standards, and professional codes of practice. While these frameworks are essential for ensuring quality and compliance, they can also introduce additional costs and administrative burdens.
- *Supply constraints*: Construction relies on three key resources: labour, plant, and materials. Skills shortages, worsened by factors such as Brexit and the COVID-19 pandemic, have reduced the availability of qualified workers. Material shortages have also increased costs and slowed project delivery. Off-site methods remain relatively underdeveloped in the UK; in 2009, modular construction accounted for only around 7% of industry output, and supplier capacity remains limited. Some progress is being made through initiatives such as partnerships between Homes England and international modular manufacturers. The impact of Brexit on the supply chain is quite dramatic and costly for the UK, the main purpose was to expand the possibility of trading globally, but in the short term it also restricted supply of materials and labour from Europe and this increased costs accordingly. For example, a development in Greenwich was using a decorative glass façade on the initial stages, but had to change the design, to a more cost-effective façade on later blocks after the Brexit/Covid Pandemic effect, partly due to cost but also due to the increased time for delivery imposed by restricted trade.
- *Public perception*: Both industry and public perceptions of modular construction have historically been negative. Industry concerns focus on cost, supplier capacity, and the lack of long-term durability data. Public attitudes are sometimes shaped by associations with post-war prefabricated housing, which was often perceived as low quality. Such perceptions can reduce demand, discourage investment in research and development, and depress market value.
- *Speculative development*: Much of the UK's residential construction is speculative, meaning properties are built before buyers are secured. In this profit-driven model, developers are incentivised to minimise costs and maximise returns, favouring familiar methods that are known to deliver acceptable margins. Significant change is unlikely unless traditional build costs become prohibitive, customers demand higher quality at the same price, or government intervention, through investment, incentives, or regulation, creates a compelling case for alternatives. If a strong financial case for modular construction can be demonstrated, however, market forces could drive its wider adoption.

- *UK construction industry overview*: The UK construction industry plays a major role in the national economy and is often seen as a benchmark for global construction performance. It is the fifth largest in the world by output, and the demand for construction is set to increase significantly in coming years. Population projections now suggest that the UK will exceed 70 million people by 2026, more than 20 years earlier than previously estimated, placing considerable pressure on the sector to deliver additional housing and infrastructure at pace.

In response to these pressures, the UK Government's Construction 2025 Strategy set ambitious targets: a 35% reduction in whole-life costs, a 50% reduction in project delivery time, and a 50% reduction in greenhouse gas emissions. Achieving these goals requires the industry to adapt, innovate, and optimise production processes across a diverse range of project types, from residential and educational buildings to commercial, leisure, retail, civil, and public sector developments. Each sector brings different client priorities, speed of delivery may be critical for educational facilities, for example, while aesthetics and customer experience are often prioritised in retail. Understanding these priorities is key when selecting construction methods and weighing up the benefits and limitations of different production approaches.

Despite the sector's importance, traditional methods remain highly labour-intensive and slow to adopt new technologies. Many innovations, such as timber trussed rafters, prefabricated concrete components, and digital tools like Computer-Aided Design (CAD) and Building Information Modelling (BIM), have taken decades to become mainstream practice. Even with government-mandated adoption for public projects in 2016, BIM is still far from universal.

Off-site construction techniques appear to be following a similar trajectory explored further in Chapters 6 and 7 Although well-established in some international markets, their adoption in the UK remains limited. In 2016, only around 7% of UK projects incorporated modular construction, and while recent surveys suggest this figure is rising, possibly to 17%, such data should be treated cautiously due to the likelihood of self-selection bias in survey responses. Growth is uneven across off-site categories, with higher uptake in sub-assemblies and panelised systems, and comparatively lower use of pod and volumetric modular approaches.

Like other sectors, construction is vulnerable to resource shortages. Skills gaps persist across many trades, exacerbated by an ageing workforce, limited recruitment of younger workers, and reduced inward migration following Brexit. At the same time, material shortages can disrupt schedules and drive up costs. Global demand fluctuations, such as large-scale infrastructure investment in other countries, can significantly reduce the availability of key materials like steel and copper, creating volatility in supply chains. Supply chains are discussed further in Chapter 4.

Another concern is the quality of finished projects. Latent defects, issues that appear after completion, remain common, leading to practices such as financial retention clauses in contracts. These clauses are designed to protect clients but reflect deeper systemic issues, including fragmented project delivery structures, inconsistent quality control, and the impacts of on-site working conditions. The multiple layers of subcontracting typical in traditional projects can also complicate warranty processes and reduce accountability.

Given these challenges, off-site construction has long been advocated as a way to improve productivity, mitigate skills and material shortages, and deliver more consistent quality. While the methods themselves are not new, their wider adoption in the UK has been slow. Internationally, mature off-site sectors demonstrate that such approaches can transform housing delivery, but achieving similar success in the UK will require overcoming structural, cultural, and market barriers.

Traditional Versus Off-Site Production

Traditional construction refers to the process of creating buildings and infrastructure primarily on-site, using labour-intensive techniques. It remains the dominant approach across much of the developed world, including the UK. These methods typically involve multiple trades working in sequence, a heavy reliance on manual labour, and exposure to external factors such as weather and site conditions.

In the UK, current building practices often prioritise high-rise developments, reduced ceiling heights, and standardised aesthetics. These design choices are frequently driven by economic pressures, including the high cost of land, the need to maximise lettable or saleable floor space, and compliance with planning restrictions. In residential master planning, the focus is often on fitting as many housing units as possible into a given area. As a result, aesthetic considerations are sometimes secondary, despite evidence that many buyers still prefer more spacious homes with traditional architectural features. Such preferences vary by demographic group and region, highlighting the need for broader, nationwide studies rather than a focus solely on densely populated areas such as London and the South East.

Traditional methods, while well-established, have been criticised for contributing to resource depletion, generating significant waste, and favouring complex, cost-driven supply chains over long-term sustainable gains. In response to challenges such as skills shortages, material scarcity, inconsistent quality, and productivity concerns, the industry is increasingly exploring more industrialised approaches. These range in scale and complexity from prefabrication and mechanisation to automation, robotics, and full reproduction.

Off-site construction (OSC) is a key example of industrialisation in the sector. It involves transferring a substantial proportion of construction activities from the site to a controlled factory environment. This shift offers several potential benefits, including improved quality control, reduced construction time on-site, and greater certainty in meeting time and budget targets. By minimising on-site labour, OSC can also reduce health and safety risks and lessen the disruption caused to local communities.

However, despite these advantages, adoption in the UK remains limited. Barriers include the construction sector's cautious culture, fragmented supply chains, and a history of mixed results in off-site projects. Legislative, regulatory, and financial uncertainties further slow investment and uptake. There is also the challenge of integrating OSC into an industry accustomed to traditional workflows and procurement models.

Studies have shown that wider use of OSC could address many of the industry's most pressing problems. Nevertheless, uptake remains relatively low, with off-site techniques accounting for only a small proportion of UK construction activity. This limited adoption is particularly notable in the housing sector, where the dominance of traditional methods persists despite growing pressures, such as the shortage of skilled labour, which may eventually necessitate a more rapid transition to modular and other off-site solutions.

UK Government Off-Site Construction Categories

The UK Government has defined the off-site construction sector through a series of categories (MHCLG, 2019). These categories group different methods according to the extent of pre-manufacturing and the type of construction element produced. The aim is to create a shared language for industry, policymakers, and educators, helping to track innovation and adoption across the sector.

- Category 1 Pre-Manufacturing: 3D Primary Structural Systems: This category covers volumetric construction, where whole three-dimensional sections of a building are manufactured in a factory and transported to site for installation. Examples include complete modules such as entire rooms or building sections, as well as partial modules like bathroom pods. Modules are often built from lightweight steel or timber frames, manufactured under controlled conditions, and rigorously quality-checked. Their engineered nature minimises movement during transport and installation, helping to ensure that the quality achieved in the factory is maintained on site. However, transporting complete modules presents challenges. Size is restricted by the dimensions of the road network, particularly in the UK where narrow routes can limit movement. Full modular systems (entire buildings) are more affected than smaller pod systems. Many manufacturers work to standardised sizes to suit transport limits; bespoke dimensions are possible but can increase costs. A further drawback is the transportation of "air", moving large, hollow modules is less carbon-efficient than transporting flat-packed or compacted materials. While volumetric construction offers high quality and faster site assembly, it must balance these gains against carbon impacts. Locating factories near raw material sources and key delivery routes can help reduce emissions by minimising overall travel distances. The principles of time, cost, and quality are explored in more detail in Chapters 4, 5 and 6.

- Category 2 Pre-Manufacturing: 2D Primary Structural Systems: Panelised systems involve manufacturing building elements such as walls, floors, ceilings, or roofs in a factory before transporting them flat-packed to site. Open panels are typically clad on one side in the factory and finished on site. Closed panels are clad on both sides, insulated, and may include windows, doors, or services installed in advance. Closed panels are more expensive but reduce site work and programme time. Panels are commonly produced from lightweight steel, timber, or structural insulated panels (SIPs). Insulation can be integrated within the panel, improving thermal performance without reducing usable floor space, a particular advantage in residential schemes where maximising internal area is important. Unlike volumetric systems, panels avoid transporting empty space, improving carbon efficiency. However, they require on-site assembly, which increases the risk of human error compared with fully finished modules. Both panelised and volumetric systems generally require cranes for installation, which introduces additional logistical and safety considerations.
- Category 3 Pre-Manufacturing: Non-Systemised Structural Components: This approach focuses on large structural elements – such as beams, columns, floor cassettes, or pile systems – manufactured in factories and then installed on site. Producing these components under controlled conditions allows for higher precision, which can reduce the size and weight of elements compared with those produced on site. These parts are typically delivered by heavy goods vehicles and lifted into place using cranes, after which construction continues in a more traditional manner.
- Category 4 Pre-Manufacturing: Additive Manufacturing: Additive manufacturing, often referred to as 3D printing, is a relatively new innovation in construction, though well established in other manufacturing sectors. The process involves depositing material – such as concrete – in a predetermined pattern using computer-controlled equipment. This method allows complex shapes to be produced directly from digital models, with minimal human intervention, increasing precision and potentially reducing waste. Entire wall sections, for example, can be printed to match bespoke designs. However, this is an emerging technology. Its long-term durability is still under review, and additional work is often required to meet aesthetic requirements. Site constraints, particularly in urban areas, may also limit the practical use of large-scale printing equipment.
- Category 5 Pre-Manufacturing: Non-Structural Assemblies and Sub-Assemblies: Many construction products are already manufactured off-site before being delivered for installation. This category includes a wide range of non-structural items, from bricks and plasterboard to

boilers and complex building services equipment. The principle is similar: manufacturing in controlled environments can produce higher-quality products, though transportation costs and logistics must be managed. A simple example is ready-mixed concrete, delivered to site in mixing lorries. On very large projects, raw materials may instead be stored on site and mixed as required.

- Category 6 Traditional Building Products with Site Labour Reduction/Productivity Improvements: This category includes factory-made products designed to speed up on-site work. Examples include pre-clad wall and roof panels (such as brick-faced systems) or pre-finished flooring sections. These systems are generally more readily accepted by contractors, as they resemble familiar products but are quicker to install.
- Category 7 Site Process-Led Labour Reduction/Productivity Improvements: Unlike the other categories, this group focuses on improving site-based processes rather than producing off-site elements. It includes technologies such as robotics, Building Information Modelling (BIM), augmented reality (AR), and virtual reality (VR). These tools can enhance productivity, improve safety, and assist in planning the integration of off-site elements into projects. For example, BIM, AR, and VR can simulate the assembly of modules or panels, helping clients, designers, and contractors visualise and coordinate works before physical installation begins.

In the UK construction industry, many projects operate under a hierarchical structure in which a main contractor oversees project delivery and manages a network of subcontractors. These subcontractors are often engaged on a short-term basis and tend to operate within local or regional boundaries. While clients and developers stand to gain significant advantages from the economies of scale that a concentrated, streamlined supply chain can offer through off-site production, traditional construction methods can also generate savings by leveraging an expansive, competitive supply chain.

One of the most widely recognised advantages of factory-based production is its ability to provide a protected, controlled, and safer working environment. Such conditions can lead to improved worker welfare, higher productivity, and reduced health and safety costs, particularly through fewer lost workdays caused by illness or injury. This greater control is achieved by concentrating activity within a fixed, well-managed production facility, rather than on an open construction site that may be spread across a large area. For example, almost half of all fatal injuries in the UK construction industry in 2017 were due to falls from height, a risk significantly reduced in modular construction, where much of the work is completed at ground level in factory conditions.

The successful delivery of off-site construction projects often depends on three interrelated factors:

- Effective management, ensuring that resources are planned, coordinated, and controlled efficiently.
- Strong communication channels, facilitating the flow of information between stakeholders.
- Robust logistics management, ensuring the timely movement of materials and components.

International examples illustrate the potential of these approaches. In Japan, off-site techniques are widely used in the mid- to high-specification housing market, while Sweden has adopted similar methods for lower-cost and social housing. In both cases, the result has been a measurable increase in productivity and uptake. These outcomes also reflect the strong correlation between innovation and overall project or business success.

Despite the recognised benefits, including shorter build times and improved quality, uptake in the UK remains comparatively low. A 2007 study found that 61% of respondents reported a neutral satisfaction rating for their use of modern methods of construction (MMC), suggesting that neither perceived benefits nor barriers were compelling enough to drive large-scale change. In the same year, further research indicated that while practitioners generally agreed on the key advantages of reduced construction time and improved quality, increased perceived costs remained a major deterrent. The discussion is further complicated by the use of multiple terms to describe similar approaches, including modern methods of construction, off-site production, and prefabrication, which can lead to inconsistencies in understanding and reporting.

Barriers and Benefits

The choice of construction method is influenced by multiple factors, with cost frequently taking precedence (Myers, 2017). However, project-specific considerations, such as client requirements, design complexity, environmental objectives, and timescales, also play a role. Each method, traditional or off-site, offers distinct advantages and disadvantages, with perceived disadvantages often acting as barriers to adoption. This section examines the primary factors influencing these decisions.

Cost

Numerous studies identify cost as the most significant barrier to off-site construction (Rahman, 2014; Mohammad et al., 2015), with uncertainty over final costs also frequently cited (Hammad et al., 2020). The UK construction market is strongly commercially driven, and cost considerations often outweigh factors such as aesthetics or end-user experience, which may hold greater weight in other cultural contexts (Liu et al., 2020). While cost is widely acknowledged as a key driver for change (Myers, 2017), it can also be both a barrier and a benefit depending on context (KPMG, 2016). This complexity is compounded by the fact that many stakeholders, including clients, contractors, subcontractors, developers, consultants, and manufacturers, may lack full visibility of the cost implications of their decisions. Later chapters address this complexity in detail, identifying over 70 cost variables, which helps to explain why research findings on the cost-effectiveness of off-site construction often diverge.

Public Perception

Public perception has a tangible impact on project design, approval, and uptake. Negative sentiment can delay planning approvals through objections, and can even influence the use of completed developments if they are boycotted or undervalued. In the UK, there is a tendency to associate faster build times with reduced quality, a perception that can disadvantage off-site methods. Cultural familiarity with traditional techniques also contributes to a preference for "tried and tested" methods, particularly given the UK's long architectural heritage. This preference persists even where modern methods could deliver better performance (Sohaib et al., 2022). Decisions are therefore often shaped by established norms rather than objective assessments of suitability. Additionally, government sensitivity to public opinion means that adoption of off-site methods may be influenced as much by societal acceptance as by technical or economic considerations (Liu et al., 2020).

Productivity

Productivity within the UK on-site construction sector lags behind other industries, with an ONS (2019) productivity index of 27 compared to 38 in manufacturing and 66 in finance. Productivity improvement, defined as the ability to produce more efficiently, depends on numerous factors, including workforce skill, site conditions, and process efficiency. A favourable working environment, with controlled temperature, safety measures, and adequate welfare provision, can improve output, while the absence of such conditions can hinder it.

The Royal Institution of Chartered Surveyors (RICS, 2024) identifies the availability of skilled workers as the most influential driver of productivity gains, followed by improvements in documentation, procurement, supply chain management, and scheduling. Interestingly, the same study suggests that prefabricated modules, equipment upgrades, enhanced supervision, and site safety improvements have relatively low perceived impact, although this may reflect entrenched industry attitudes rather than empirical performance.

Off-site production, undertaken in controlled factory environments, has the potential to raise productivity by reducing on-site construction time. It can also improve health and safety by limiting workers' exposure to hazardous site activities. However, industry focus in the UK remains heavily oriented towards expanding the on-site workforce rather than shifting to innovative production methods.

Environmental Considerations

In addition to cost, environmental impact is an increasingly important factor in construction decision-making. This is commonly framed through the "three pillars" of sustainability, economic, social, and environmental, which must be balanced to achieve sustainable development. In practice, trade-offs are common – for example, reducing costs can sometimes conflict with reducing energy consumption.

From an environmental standpoint, construction's primary goals include conserving resources and reducing carbon emissions. This applies both to operational energy (during a building's use) and to embodied energy and emissions during construction. Transport of materials is a major contributor to construction's carbon footprint. Traditional on-site methods require the delivery of multiple raw and manufactured materials to temporary sites, often over long distances. This includes not only the transport of finished products but also the transport of raw materials to manufacturers.

Off-site manufacturing can mitigate this by locating production facilities close to raw material sources, reducing transport distances and associated emissions. For example, a brickworks situated near a clay source minimises haulage requirements. However, certain off-site approaches, particularly modular construction, face transport-related inefficiencies. Moving large, fully formed 3D modules often involves transporting significant unused space ("air"), requiring more trips or larger vehicles than flat-packed materials. Panelised systems can be more transport-efficient, delivering components in a compact form.

Environmental performance also depends on the building's thermal efficiency, measured as its U-value. Lower U-values indicate better insulation and reduced heating or cooling demand. Traditional brick-and-block systems typically use a wide cavity with insulation, which can reduce usable floor space. Timber-framed or panelised systems can achieve the same insulation in a thinner profile, improving space efficiency. Moreover, the precision of factory-built components reduces the risk of gaps, thermal bridging, and other performance issues that can arise in uncontrolled on-site environments.

While renewable technologies (e.g., solar panels) can further reduce operational energy use, these are independent of the choice between traditional and off-site methods. Instead, the construction method influences the quality and efficiency of the building fabric itself, which is a critical determinant of long-term environmental performance.

Material selection is very important in the reduction of carbon footprint – most contractors now carry out environmental impact assessments on projects which will include evaluation of carbon footprint (how much carbon is used in the extraction, manufacture, delivery and installation of a material). Environmental Product Declaration (EPDs) are attached to products showing this linked to life cycle assessment, which evaluates the material's use through its life. In addition, successive governments develop and enact legislation and codes of practice to ensure the appropriate selection of materials such as PAS2080, for example, and it is worthwhile to explore this topic further.

Waste Reduction

Waste generated on construction sites poses significant environmental risks, health and safety hazards, and can contribute substantially to project costs. Therefore, effective waste management is crucial. Waste arises from various sources and takes many forms, making its management complex but also presenting multiple opportunities to improve sustainability. Broadly, waste management can be divided into two categories: waste disposal and waste production reduction.

Proper disposal is essential, and a well-organised, multilayered skip system is recommended to ensure waste materials are stored appropriately before disposal. Wherever possible, waste should be recycled, reused, or repurposed. For example, unused materials can be returned to suppliers, transferred to subsequent projects, donated to local community initiatives or educational institutions, or sold through yard sales. Such practices not only reduce landfill use but also generate cost savings.

Reducing waste production from the outset is equally important. Building Information Modelling (BIM) can help by providing precise measurements and enabling the use of standardised components, thereby minimising waste. Consider plasterboard: standard sheets measure 2.4 metres in height, whereas the common UK ceiling height is 2.3 metres, leading to inevitable offcuts. Standardising component sizes, often achievable with off-site construction techniques, can help eliminate such material wastage.

Waste generated during factory production must also be considered. Transferring waste from site to factory without additional benefit offers little value. However, the controlled environment of a factory allows for waste reduction at the source. For instance, timber can be ordered in standard lengths to reduce excess, and strict quality control can prevent accidental waste caused by production errors. Fixed factory locations also facilitate more efficient waste disposal, such as recycling timber waste for heating, and enable partnerships with local communities to promote sustainability initiatives.

Time Impact

Modular construction methods have been demonstrated to reduce on-site production time, which in turn enhances on-site productivity. As such, technological advances that boost productivity invariably influence overall costs. The interval between groundbreaking and project completion shortens when modular techniques are employed, particularly when the substructure is also prefabricated off-site (Kamali and Hewage, 2016).

A significant challenge of adopting modular construction is the requirement for earlier order placement compared to traditional on-site methods (Goulding et al., 2015). Additionally, emerging supply chains may experience fluctuating lead times due to demand cycles: as demand grows, supply chains stretch and delivery times increase until new production lines are added, after which lead times shorten again. This cyclical variability complicates project planning and impacts costs. Early payment for materials is often required in modular builds, contrasting with traditional credit arrangements where payments might follow project milestones or sales.

Early, detailed pre-project planning is critical, as module designs must incorporate layout, build processes, transportation, installation, and connectivity considerations well in advance. Consequently, last-minute client changes are difficult, if not impossible, to accommodate without incurring additional costs or affecting market value.

Resource Shortages

One driver behind shifting construction methods is the shortage of skilled labour and materials. Manufacturing environments generally demand lower-skilled labour than on-site construction because of their controlled, automated processes, which reduce the need for complex problem-solving (Goulding and Rahimian, 2020). Research predicts a forthcoming skills crisis in the construction sector (Farmer, 2016), necessitating either increased recruitment or adoption of methods that reduce reliance on highly skilled trades, such as off-site construction.

The UK's departure from the European Union has intensified labour concerns, with approximately 9% of construction workers originating from the EU, and in London, this figure rises to 54%. Attracting new entrants remains difficult due to low interest among school leavers. Off-site construction, whether domestic or imported, offers a viable response. Higher-level skills related to planning and Integrated Project Delivery (IPD) systems remain vital to facilitate successful off-site implementation.

Modular construction reduces the need for on-site trades and the coordination effort required to manage them. Craft-level workers are typically slower to adapt to new methods and technologies.

A dysfunctional education and training system exacerbates sector issues; without improvements here, challenges will persist regardless of construction methods adopted. Fixed-location modular production sites also enable the development of sustainable local labour communities. Unlike the mobile workforce of traditional construction, a stable manufacturing plant can be strategically placed in areas of high unemployment and low labour costs, promoting community regeneration and greater operational efficiency.

Material shortages pose another challenge. Construction's project-by-project nature leads to irregular material orders and a fragmented supply chain. Conversely, a fixed factory setting supports stronger, more consistent supply chains less vulnerable to demand fluctuations. Thus, off-site production not only addresses labour shortages but also strengthens material supply resilience.

Quality

Traditional on-site construction has often been criticised for failing to meet required quality standards, prompting calls within the industry for enhanced quality control to deliver better-quality homes. One of the most significant costs in construction arises from rework due to poor quality, errors, or damage. Such additional tasks, whether caused by incorrect installation or defects identified later, result in unplanned labour, material, and time expenses. In some cases, rectifying these issues can cost as much as or more than the original work, particularly when dismantling and rebuilding are necessary.

Even more costly are defects that only emerge post-completion, such as a faulty door frame discovered after occupation. Rectifying these problems often requires site revisits, disrupting other projects and potentially damaging adjacent finishes, thereby increasing cost and inconvenience.

Despite their impact, these extra costs frequently go unrecorded, as subcontractors often rectify mistakes without formally tracking expenses, absorbing losses into overheads. However, fewer errors improve profit margins, which over time can lead to more competitive pricing, ultimately benefiting clients, even if these savings are not explicitly documented.

Off-site construction delivers superior quality and precision, owing to the stringent Quality Assurance (QA) processes enabled by factory production, which exceed those possible on-site. This is further supported by higher levels of supervision and process monitoring in controlled environments. Because modular units must be transported and lifted into place, materials used are durable, lightweight, and weather-resistant. The controlled factory environment protects materials, allowing higher-quality finishes than typically achievable on site.

Greater standardisation of components and measurements enhances delivery reliability, quality, and cost predictability for construction projects. The adoption of Design for Manufacture and Assembly (DfMA) principles further improves these outcomes. Increased quality leads to reduced rework, contributing to cost savings.

This is explored much further in Chapter 6, but one key consideration is the installation of the MMC product, as while in factory production should be of higher quality, when the item is installed on site the skill of the installation team can impact on the finished product. Also there are usually on-site elements that need to be installed around the MMC items, often attached to, and once again this brings in an on-site, reduced quality aspect that needs to be allowed for. Lack of knowledge of the product will increase this potential

Design Flexibility

A key barrier to off-site and modular construction adoption is perceived lack of design flexibility. This stems from two main factors. First, modules must comply with transportation restrictions via highway systems, limiting size and shape. Second, production orders must be placed well in advance to align with manufacturing schedules, necessitating early design freeze.

This early commitment to design requires developers to develop new skills for adjusting work breakdown structures and project programming to improve planning accuracy. Supply chain management must be carefully considered, as modular construction offers less scope for alterations or reactive changes during the build compared to traditional methods. For example, all internal design and finishes must be planned and ordered early to accommodate factory production and delivery times, whereas traditional construction allows reasonably straightforward sourcing and implementation of late changes.

This inflexibility can deter some clients, although it is common practice in speculative development. Furthermore, communication methods need to evolve beyond those used in traditional construction, due to the specific requirements of the main modular supplier, installation contractors, and associated subcontractors or suppliers.

Other Barriers and Benefits

The size and maturity of the supply chain is a significant factor influencing the choice between off-site and traditional methods. If off-site systems are difficult to source due to a limited supply chain, contractors and clients may opt for conventional on-site construction. While traditional builds may experience fragmented communication and supply chains, modular construction demands closer, stronger partnerships.

Some organisations have attempted to develop their own production facilities, but the high upfront costs have limited success and, in some cases, caused business failure.

The construction sector is a complex network of organisations with differing aims, making sector-wide change challenging. However, pressures to innovate and reduce costs are driving change. Without adoption of new methods, the UK is projected to face insufficient production capacity to meet built environment demands by 2025. Consequently, the UK government has shown increasing commitment to modern methods of construction.

Off-site construction prevents multiple trades from congesting the same space and time, improving efficiency. This counters the law of diminishing returns often experienced on traditional sites, while enabling greater workflow continuity and process control. Organisational learning helps companies adapt and survive within the evolving industry landscape.

Reducing supply chain size has both positive and negative effects. A smaller supply chain increases reliance on each supplier's success; failure of a single organisation can have greater impact than in traditional, more diversified supply chains. Traditional construction can often more easily replace subcontractors or suppliers if issues arise. Conversely, a smaller, specialised supply chain facilitates stronger relationship development and longer-term partnerships, which can benefit all parties involved.

Concluding Thoughts

The book is not aimed at making the case for the use of MMC on construction projects, it is not trying to present a business case advocating the use. It will not present costs and show the financial benefit for your project – the cost variables are too varied and each site's uniqueness, including the individual needs of the client, make this an almost impossible task. What the book does do is explore the considerations around the use of MMC – it will look at the good and the bad the potential for increased quality and time on the projects, the concepts of costing and pricing mechanisms. It will explore how these can work well, and how they can come unstuck when using MMC. And within he pages we also aim to show techniques and process that can be established on projects to help in making them work, so that the MMC benefits, if needed, can be maximised.

References

Farmer, M. (2016). *Modernise or Die: The Farmer Review of the UK Construction Labour Model*. London: Construction Leadership Council.

Goulding, J. and Rahimian, F. P. (2020). *Offsite Production and Manufacturing for Innovative Construction. People, Process and Technology*. Oxford: Routledge.

Goulding, J. S., Rahimian, F. P., Arif, M., and Sharp, M. D. (2015). New offsite production and business models in construction priorities for the future research agenda. *Architectural Engineering and Design Management*, 11(3): 163084.

Hammad, A. W. A., Sutrisna, M., and Zaman, A. (2020). Towards optimising OSM implementation in construction projects: A sustainable procurement perspective. In Goulding, J. and Rahimian, F. P., *Offsite*

Production and Manufacturing for Innovative Construction. People, Process and Technology. Oxford: Routledge.
Kamali, M., and Hewage, K. (2016). Life Cycle performance of modular buildings: A critical review. *Renewable and Sustainable Energy Reviews*, 62: 1171–1183.
KPMG (2016). *Smart Construction: How Offsite Manufacturing Can Transform Our Industry*. London: KPMG.
Liu, Y., Mao, C., and Wang, X. (2020). Offsite construction driving forces: A critical reflection of Hofstede's national culture model. In Goulding, J. and Rahimian, F. P., *Offsite Production and Manufacturing for Innovative Construction. People, Process and Technology*. Oxford: Routledge.
MHCLG (2019). *Modern Methods of Construction: Introducing the MMC definition Framework*. London: MHCLG.
MHCLG (2025). Subnational estimates of dwellings and households by tenure, England, quality and methods guide. www.ons.gov.uk/peoplepopulationandcommunity/housing/methodologies/subnationalestimatesofdwellingsandhouseholdsbytenureenglandqualityandmethodsguide. Accessed on, 14th September 2025.
Mohammad, F. M., Baharin, A. S., Musa, M. F., and Yusof, M. R. (2015). The potential application of IBS modular system in the construction of housing scheme in Malaysia. *Procedia Social and Behavioural Sciences*, 222: 75–82.
Myers, D. (2017). *Construction Economics: A New Approach*. New York: Routledge.
ONS (2019). Construction statistics: Number 19. www.ons.gov.uk/businessindustryandtrade/constructionindustry/articles/constructionstatistics/number192018edition (accessed 12 October 2019).
Rahman, M. M. (2014). Barriers of implementing modern methods of construction. *Journal of management in engineering*, Jan/Feb issue pp 69–78.
RICS (2024). *RICS Construction Productivity Report 2024*. London: RICS.
Sohaib, M., Tengyue, H., Yu, Q., Sayed, K., and Ruodan, S. (2022). How a successful implementation and sustainable growth of e-commerce can be achieved in developing countries: A pathway towards green economy. *Frontiers of Environmental Science*.
Towey, D. (2018). *Construction Quantity Surveying: A Practical Guide for the Contractor's QS*, 2nd edn. New York: John Wiley and Sons.

Chapter 2

Modern Methods of Construction

Introduction

Modern Methods of Construction covers a wide range of possibilities and has been used for a number of years. The basic concept is that parts of a building or civil engineering project are manufactured off-site and delivered to site for installation. Using this definition alone almost every item used in a modern construction project would fall into this category – bricks are manufactured in a factory and delivered to site, electrical switches are manufactured in a factory and delivered to site for installation, etc. – therefore it is important to be a bit more selective in what is termed Modern Methods of Construction (MMC) or off-site. This chapter address these concepts. While exploring the variety of MMC available to the industry today there are a lot of aspects that need to be taken into account. For example, what is the production process and how does this impact on our roles as cost managers or project managers on a construction site? What are the various drawbacks and benefits that relate to this type of production method? These questions plus others are all discussed below, some in detail and others as a feeder into later modules.

Modern methods of Construction (MMC)

Definition

The term modern methods of construction is a bit of a misnomer as the process included in a normal version of this are not really modern as they have been around for a long time. For example, timber framing has been used as a concept in the industry for over 100 years, and volumetric was used following the second world war as a means of providing adequate and quick housing following the devastating effects of that war. Therefore, other terms may be more relevant to categories the work being done in this sector of the industry.

Perhaps off-site construction explains the processes better than MMC? But as suggested above, this is too wide ranging as most of what is installed in our contemporary projects is manufactured off-site.

In 2019, the UK government introduced a framework designed to bring consistency and structure to the way Modern Methods of Construction (MMC) are defined and applied across the industry (MHCLG, 2019). Prior to this, the lack of shared terminology often led to ambiguity between clients, designers, and contractors. The framework has therefore become the principal standard in the UK, offering a coherent system that supports benchmarking, procurement, and policy alignment. It has been widely adopted because it enables stakeholders to communicate more effectively, measure performance with greater accuracy, and embed innovation in a structured way.

DOI: 10.1201/9781003570431-2

At the core of the framework are seven categories of MMC. These capture the wide range of practices being used to modernise the industry, spanning pre-manufacturing techniques, process optimisation, and digital technologies. While not all categories relate directly to off-site construction, the framework demonstrates how innovation can be integrated throughout the project lifecycle. Its strength lies in providing a flexible yet standardised approach, enabling different project types to be compared and evaluated consistently.

The importance of the framework is reinforced by tools and guidance developed around it. For example, the NHS MMC Assessment Tool applies the seven categories as a means of evaluating projects, ensuring a consistent approach to assessing compliance and encouraging the early adoption of modern methods. Similarly, the government's Construction Playbook (Cabinet Office, 2020) explicitly references the framework, setting out how public sector procurement should drive innovation by embedding MMC within contracts and supply chains. Both of these documents demonstrate the framework's role as a practical mechanism for implementation, not just a theoretical model.

Overall, the 2019 framework has become a cornerstone of UK construction policy, shaping how MMC is defined, applied, and assessed across both public and private sectors.

Category 1 Pre-manufacturing (3D Primary Structural Systems)

The three-dimensional units used in this system are manufactured in a factory and can be used to replace part of a building project or entire sections or even the entire project. Take, for example, the use of bathroom pods – an entire bathroom including its walls is manufactured complete with all fittings in a factory environment, delivered to site and then craned into position while the building is being constructed. The bathroom pod is then connected to the relevant services and fixed to the structure but is then ready to use. This saves the contractor all of the work and trades needed to install the bathroom on site, including, for example, plumbers and electricians for the services, plasterers and decorators for the surface coating, carpenters to construction the walls, and tilers for the finishes.

The other aspects of completing an entire section of a building can be seen with the example of a hotel, where the entire bedroom unit is manufactured in a factory and then delivered to site and, again, craned into position – the corridors, stairs and lift shafts are typically constructed on site, typically using an in-situ concrete structure. The modular three-dimensional units are then installed and connected to the services and each other. Sometimes the modules will have the cladding installed in the factory and sometimes installed on site. But the bulk of trades that would normally operate on the construction site are replaced by the operation in the factory environment.

The framework stipulates four alternate sub-categories for this type of system. Namely:

- Structural chassis only – not fitted out: This is where the basic skeleton of the 3-dimensional unit is manufactured in the factory but all finishings, fittings and services are installed on-site
- Structural chassis and fit out: This is similar to the second example given above, where all services and internal fit out is completed off-site, but the external cladding is completed on site
- Structural chassis, fit out and external cladding/roofing complete: This is completing all the work needed for the project in a factory environment except for foundations, groundwork and landscaping, and some connectivity work.
- Structural chassis and internal fit out – "podded" room assemblies – bathroom/kitchens, etc.: This is similar to the example given above for bathroom pods, where "smaller" sections of the build are completed and installed within the main build.

While these categories are clear, industry seldom follows this principle, and often the lines between categories are blurred. For example, getting a structural chassis, with partial fit out. As the supply chain is in its infancy, often manufacturers will move from manufacturing norms and provide more bespoke solutions like this, even though it may go against the standardized principles that make MMC more beneficial.

Category 2 Pre-manufacturing (2D Primary Structural Systems)

This category is one of the most common forms of off-site construction used in the UK currently. It is a 2-dimensional panel system made from a variety of materials, manufactured to size and to a variety of finishes in a factory setting and then transported to site. There are some key differences between this system and that of the 3-dimensional system often given as benefits and addressed in more detail below. As the panels are 2-dimensional they can easily be stacked and transported as one, often large projects can be delivered in one lorry load, this is not the case for the 3-dimensional modules above as they are transported ready assembled. One of the criticisms of this is the fact that you are paying to transport air in the modular system, as well as the vehicular based emissions, that is not the case for the panelised system. Also, another presented advantage is, depending on the size of the panel and the location it is being installed, it may be possible to manoeuvre the panels using manual labour without the need for mechanized assistance such as forklifts or cranes. Again, this is not possible with the 3-dimensional systems. However, one key disadvantage is that with a panelised system there is more requirement for onsite labour than there is with the 3-dimensional systems, and, as can be seen below, a big advantage of using off-site production methods is the reduction in on-site labour – panelised systems can reduce this benefit.

The typical materials and finishes used in panelised systems is quite varied but will typically be a framed structure using timber or metal framing with some form of boarding finish to the outside and inside surface. Typically, OSB board will be used on the outside due to its weather proofing nature, but with new technologies and advancements in material manufacturing this is a constantly changing preference. On the inside it is typically some form of gypsum-based board but again this is a constantly changing environment and so needs to be monitored and updated as it develops. In the skeleton structure panelised system, insulating material such as cellulose or fibreglass, or sheep's wool, is placed in the gaps between structural members.

The other standard form of panelised system is a structural insulated panel system (SIPs). In the skeleton structure panels above, the strength and load transferring ability of the panel comes via the structural skeleton, mainly the upright members. In a SIPs panel the structural element is the boards used on the surfaces, and the entire inner core is made up of a cellulose type material. One of the drawbacks of the skeleton panel system is that where the structural members are located creates a potential for a small cold bridge, whereas by filling the entire inner core with the insulating material the thermal performance of this system is improved.

Another group of modern panelised systems uses cross-laminated timber (CLT), light-gauge steel framing (LGSF), and hybrid composites. CLT panels are formed by layering timber boards at right angles, creating strength in both directions, while LGSF relies on thin, cold-formed steel sections assembled into lightweight frames. Hybrid composites combine different materials, such as fibre-reinforced polymers with timber or steel, to achieve both strength and reduced weight.

A key consideration for these systems is fire safety, which is assessed using the European classification standard BS EN 13501. Materials are tested for their reaction to fire and resistance to fire, with performance ratings that influence their permitted use in buildings of different heights and occupancies. This is particularly significant for CLT, where the behaviour of structural timber in a fire remains the subject of ongoing research and debate. While engineered timber can char in a predictable manner, helping maintain structural stability, concerns remain around concealed fires, smoke spread, and long-term durability in high-risk environments.

These issues link directly to the Building Safety Act (2022), which applies stricter requirements to higher-risk buildings, including many multi-storey residential schemes. Where CLT or other combustible materials are proposed for such projects, designers and contractors must demonstrate robust compliance with fire-safety regulations and ensure that the material choice supports both safety and regulatory approval. The advantage of these systems, however, is that they offer opportunities to balance structural performance with sustainability, enabling innovative solutions that can meet both environmental and safety demands when correctly designed and regulated.

Another thing to note, is with modern advancements it is possible to have many of the structural elements of a project such as walls, floors, roofs, etc. built using the same systems. This will assist in providing the advantages over a much larger section of the project. To sub-categorize the discussion above, the framework has created three for ease as seen below:

1 Basic framing (only including walls, floors, stairs and roofs): This is provided with just the structural elements as listed above without any of the finishes or insulation.
2 Enhanced consolidation (insulation, internal linings, etc.): This system is as explained above where the panels come partially completed with boarding installed but requiring other aspects to be completed on site.
3 Further enhanced consolidation (insultation linings, external cladding, roofs, doors, windows): As can be seen from the list much more of the finished work, especially on the external aspect of the build is completed in the factory setting, gaining more of the benefits as listed below.

Adding more factory installed items will enhance the benefits (discussed more below) such as quality and time, while the costs benefit still needs to be analysed (see Chapter 7). It is also worth noting that while transportation is easier and considered more environmentally friendly with panelised systems, there are still access challenges associated with site access, delivery routes and installation (see Chapter 4 for more detail)

Category 3 Pre-manufacturing Components (Non-systemised Primary Structure)

Category 3 is made up of components of structural elements manufactured off-site and used as part of the structural element of a building. These can be sub-structural elements such as precast steel, concrete or timber piling systems used to support heavy-weight structures or in low strength soils. The piles would be driven into the soils using mechanized systems such as pile drivers, and as such are not suitable in every location.

There are also steel, pre-cast concrete or timber product items such as pile caps, ring beams columns, and beams used in providing the skeleton support structure of a building, for example, in the manufacture of portal frame buildings used in many factory and warehouse projects. Precast flooring systems and slabs also falls into this category while there is also a potential for floor slab panels to also fit into Category 2, a natural split would be the size of the system.

Typically, the smaller precast concrete systems would be in Category 3, but the larger more lightweight shorter span timber system would be in the panelised Category 2.

Pre-assembled timber, steel, and precast concrete staircases would also be part of this category – these are a common element in many house builds and not often considered as off-site construction, but it is rare in modern projects for any staircase to be cast on site or manufactured on site, as it is quite a complicated process to ensure the correct elevation is created with the riser height, and also that the correct travel is created with the correct going height. Building regulation K is also quite strict in terms of what is acceptable or not in staircase production, as it is a place where many accidents can happen and can often cause severe injury or even loss of life if they are incorrect. Therefore, most project managers will opt for the additional control and quality that comes from a factory set manufacturing process when it comes to staircases.

Recent developments in building regulations, such as Part L (Energy Efficiency) and Part O (Overheating) have impacted on panel design, and it is important to have very early cross collaboration between project and panel design teams to ensure these are considered throughout. It is also worth pointing out that there are many different parts of legislation that impact on construction projects, and the more complex the project often the more complex the legislation – it is therefore imperative that all stakeholders in the industry keep abreast of changes, and a great deal of time spent as a professional in the industry should be keeping yourself updated. The procurement of components is covered more in Chapter 3 and this should be cross-referenced against the developments not only in legislation, but also to codes of practice, best practice, benchmarking, etc. – for example, explore the construction playbook.

Another commonly used aspect that would fit in Category 3 is that of the pre-manufactured roof truss system, typically made from timber or timber-based products – this is the most common form of roofing used in domestic housing production, and is a system that has now been used as the most popular choice for over 30 years. A question discussed later in the book is why industry is reluctant to use off-site construction techniques when the benefits of using them in situations such as roof spans and staircases has been the case and not questioned for a number of years.

Category 4 Additive Manufacturing (Structural and Non-structural)

3D printed products fall into this category and this can cover entire buildings which is possible with current technologies, or it could be any number of components used within the production. While this is a reasonably new technique, there are only certain situations where this can be used effectively and while the system developers and manufacturers continue to come up with new and innovative ways to use this process, it will probably take a number of years before it becomes more mainstream. There are some advantages to this system, though, as there is limited transportation needed, such as was discussed as a key drawback for the Category 1 and 2 above, the 3D printing tends to happen on site and so once the plant and equipment is in place, there is a need to transport the printing material used, but other than that the main structure at least is manufactured and installed in-situ. Another key advantage is the machine does the bulk of the work with minimal supervision, again removing one of the key drawbacks of site work, that of human interaction and potential error. Also, the intricacy of the design is possible due to the process and printing the walls in layers – this is not always as possible with the Category 1 and 2 above, without significant cost increase.

The key drawback to this system currently is the limitations in use as the process is in its infancy and therefore is only of use in limited applications. Large-scale 3D printing – where wall sections, etc. are printed using a fibrous concrete mix – is still experimental. The 3D printing of smaller components can be incorporated reasonably easily, and it is likely to be this area that develops first, while large scale may or may not follow later as the system increases in accuracy, application, and cost.

Category 5 – Pre-manufacturing (Non-structural Assemblies and Sub-assemblies)

There does appear to be a crossover in elements of this category and Category 1 and 2 above, but the key difference is that products and processes in Category 5 are non-structural, whereas those in Category 1 and 2 are structural and load bearing. Therefore, one of the elements in Category 5 is listed as bathroom or kitchen pods (also listed in Category 1), but in Category 1 the walls, floors and roofs of the pods are part of the structure of the building, whereas here, the pods are installed inside a structure and therefore only carry their own load. As well as pods used for bathrooms, water closets, kitchens and the like, pods can also be used for other uses, such as storage, or mechanical and electrical equipment enclosures, etc. – typically, any room that can be completed off-site, but does not need to be a part of the structural element of the building could fall into this category.

Another area where off-site construction has seen growing use in recent years is in Mechanical, Electrical, and Public Health (MEP) systems. In this approach, entire components or sections of services systems are manufactured in a factory environment and then delivered to the site for installation. This reduces the need for complex on-site assembly and minimises the risk of errors or delays.

The potential for enhanced quality is a key advantage of factory-produced MEP systems. Components can be assembled and tested under controlled conditions, allowing them to achieve necessary certifications, including fire performance standards such as BS 8414. The benefits of this approach have also been extended to complete façade systems, particularly in response to the Grenfell Tower disaster. Testing and certifying complete systems in a factory setting ensures that the components work together safely and consistently, rather than relying on the combination of separate systems being completed on site.

Most of these discussed so far are reasonably new developments in the industry and are most commonly used on larger projects where off-site construction benefits can better be realized. As with other areas of modern construction, the adoption of Building Information Modelling (BIM) is supporting this process. Accurate digital models allow components to be manufactured to precise specifications and installed without the need for adjustment on site. This ensures that the final system fits correctly within the building and performs as intended, improving both efficiency and safety.

By moving production into a controlled environment and integrating design through BIM, off-site MEP and façade systems, etc., offer improvements in quality, safety, and reliability, while also reducing programme risk and the potential for on-site errors.

There are two aspects in this category however, that, similar to roof trusses and staircases discussed in Category 3 above, are quite common and have been used for a while. Again, suggesting the case for off-site is stronger than currently realized. The first of these is façade assemblies – for example, composite walling used on many factory and warehouse projects. This also includes curtain walling, glazed or panelised systems used on most high rise and larger projects as a means of covering large areas quickly. These panels are completed off-site in factory conditions and transported to site and installed onto the existing structure.

The second is the use of door cassettes. In traditional construction systems a joiner would install a door lining, hang a door, install the ironmongery (door handles, locks, etc.) and then finish the trim, often with an architrave. A door set system has all of these elements installed in factory conditions, and transported to site and installed in one go. It makes the accuracy on site much better, and the whole process is much quicker with better quality control. Again, this shows that off-site construction can work and is working in many cases.

Category 6 Traditional Building Product-Led Site Labour Reduction/Productivity Improvements

Category 6 looks at systems installed on site but are generally in a larger format than those used in traditional building. For example, this can include the use of larger format blocks used in walling, brick slips systems for external cladding. Generally, it covers new systems that enhance the speed of on-site installation. Therefore, not gaining the full advantage of off-site that can be realized in Category 1 and 2, for example, but still reducing some of the issues that can be seen with on-site production. It would also include cut to size elements, thereby reducing another drawback to onsite production, that of waste.

This reduces the need for highly-skilled and costly labour on site, improving accuracy, as it removes the human risk element, which helps to counter the cost of the equipment needed. As with many equipment-led savings, there is a tipping point achieved whereby the more a costly piece of equipment is used, the more financially beneficial it becomes. The more designers utilize large-format systems requiring installation equipment, rather than human skill, the more cost-effective these systems will become (Myers, 2017).

Category 7 Process-Led Site Labour Reduction/Productivity/ Assurance Improvements

The final category does not use off-site as part of its manufacturing process, but instead looks at ways to minimize the issues seen with traditional production – for example, completely covering the building site in a weather protection system, allowing productivity to continue without the impact and delays caused by weather.

Another aspect covered in Category 7 is the use of faster support systems, such as modular scaffolding needing less time to erect and alter on site than tubular scaffolding systems; or slip form formwork systems that allow for quicker in-situ concreting; tunnel formwork systems that allow for continuous tunnel installation; or insulated formwork that control the speed of curing for a better product finish – basically, systems that improve on-site quality and speed but still use on-site labour and processes.

Building Information Modelling also falls into Category 7 as do other lean productivity systems and processes – creating more systemized and digitally enhanced productivity on site, with fewer errors and slicker processes. Included with this digital enhancement of on-site working would be the use of VR and AR. This is where workers on site can use BIM models to visualize the areas they are working in as they work – for example, an electrician wearing digitized visual aids can "see" the location of light switches without the need to measure. Added to this is the tool enhancement that will not allow installation of elements in the wrong place – for example, a drill that will only allow the operative to drill the hole where the BIM drawing states the hole should be, anywhere else and it won't work. Or plant that follows the levels or dimensions set out in the BIM drawings when excavating foundations, without the need for engineers to do complicated setting out procedures.

This leads into another area covered in this category of worker planning wearables, such as GPS and health monitoring equipment, sold to workers as a means of keeping them safe and monitoring their health and well-being while working, but which can also be used as a productivity tracking device. Exoskeletons are another system used to not only improve worker safety but can also to enhance productivity as workers are able to safely lift heavier loads.

Finally, in this category is the use of non-manual labour elements such as robotics and drones – used again to monitor productive and planning and also to improve productivity by removing the human error or frailties and using mechanical systems to replace this aspect.

As with 3D printing discussed in Category 5, many of these systems are in their infancy but some are being developed quicker than others due to their having more impact on processes – such as the use of drones in monitoring work quickly and safely, and for assessing productivity. Quantity surveyors are now able to remotely monitor work flow when developing their valuation claims, or assessing the accuracy of claims, without the need to even visit the site. Construction Excellence has developed a Value Toolkit, that is utilised to enable beneficial decision making (Excellence, 2022). Other initiatives can also be used, such as CITB Digital Toolkit to help companies develop digital skills within their projects. All of these innovations are jointly driven by client/industry desire, and government drive/legislation (see Chapters 4 and 6 for more details).

Production Processes

To get a full appreciation of the uses, benefits and drawbacks of the various MMC categories, it is important to appreciate the way in which these products are manufactured. One of the key drawbacks to using on-site labour is the lack of quality control that is possible (see Chapter 6). Moving the production into a factory environment should provide more control and therefore greater levels of quality. There is, however, an issue when the product then needs to be transported to site for installation – therefore stresses and loads need to be considered not only for the building itself, but also for the transportation and installation, and the stresses and strains that can be placed on the item in question.

On construction sites, visibility of the production process is difficult and as the project progresses this becomes even harder. Typically, on a construction site the control comes from the site management team and they are often based in a fixed site office, with limited views over the site. On large sprawling sites, such as civil engineering projects or housing estates, the site office will only have direct visibility of the immediate surrounding area – to get a more complete view of worker progress the site office would either need to be moved several times as the site develops, which is expensive, or the site management team will need to spend more time out of the office and around the site. In contemporary projects, this is not always possible, due to the amount of paperwork associated with running a site. On more compact sites, such as inner-city projects, this becomes a problem, in that the positioning of the site office becomes more about not getting in the way of the building than it does about visibility. Once out of the ground, high rise buildings rise fast and as soon as the external wall closest to the site office goes up, the visibility of the site has gone. It is possible to use technology to monitor and control – such as CCTV, personnel monitoring devices and drones – but this type of monitoring system is in its infancy in the industry and there is a reluctance to use it.

When working in factory conditions, the positioning of the office, the open plan layout typically used and the more controlled working environment, makes the control and monitoring of working practices much easier to manage. This will ultimately lead not only to more efficient production but also greater quality of output. This principle can be applied to many of the construction aspects used today, thereby improving on overall quality.

For Those Wanting to Work in MMC – Things to Consider

This book is mainly aimed at those wanting to work in the construction industry but from a site and installation perspective. With the growing use of MMC, there will also be opportunities for graduates and experienced professionals to work for MMC manufacturing companies as the need for greater industry knowledge and costing expertise increases. There are some considerations to take into account when making the move from a predominantly site-based knowledge and expertise into a factory-based production facility.

Location

Typically, steel or timber is the predominant material in contemporary off-site products, although this is not a restriction and many other materials can be used. From an environmental, and carbon footprint aspect, the location of the main production facility is key. This is also important from a cost perspective. For example, locating the factory and production facility close to a source of material will reduce the transportation time of getting that material from source to factory. This will not only reduce costs, it will also reduce harmful emissions used in the typical transportation of those materials. While there are some good developments in the transportation industry in trying to address this with more electric vehicles, and the development of hydrogen fuel vehicle, this is still a big consideration.

The majority of timber that is of sufficient quality to be used in construction is imported, from Canada or Baltic countries such as Latvia. Therefore, locating the factory close to a port would be sensible with the use of timber. Steel and other metal products are also mainly imported – although there still remains a strong steel production element in South Wales and some in the East Midlands, the steel industry is not the same as it was 50 years ago in the UK. Therefore, locating the factory close to a steel plant, such as the large facility in South Wales or East Midlands, or again close to a port would also make sense.

The key ports in the UK for importing materials are

- Port of Felixstowe
- Port of Southampton
- London Gateway Port
- Liverpool Port
- Immingham Port
- Dover Port
- Port of Milford Haven

Labour prices are also a factor to consider when locating factories, as some of the major ports are also located in areas of low employment and low employment competition leads to low employment costs.

On-site versus Off-site Techniques

There is an ethos when working on site, developed over many years, that allows for a certain amount of flexibility and on-site adjustment. And while this allows for last minute changes and the use of high-level skills, it also leads to the risk of lower quality and errors. The use of MMC should be different, and, where MMC works well, on-site adjustments should not be needed or allowed as once this happens the benefit of off-site production disappear. An example of this would be if a timber frame house is manufactured to plan, but when the groundworkers are installing the sub-structure slight changes are made so that the panels no longer fit. This then requires on-site adjustments of the panels and all of the advantages of manufacturing in controlled conditions are no longer benefits.

Therefore, to maintain the benefits of off-site production it is important that all on-site activities are built exactly as designed – this is an aspect of the industry that has improved in recent years due to the increased use of Building Information Modelling (BIM). While this is discussed in more detail in later chapters, the key principle of BIM is that the finished product and all of its constituent parts are designed digitally to be in their optimal position. The designed elements are then installed in exactly that location when on site. Thereby, if the installation matches the optimum design, then the installation becomes more accurate.

MMC is an ideal solution to work with BIM, as the ethos that should be used in MMC production is more closely aligned with manufacturing than it is with construction. It is very important when working with standardised products, such as those produced using off-site methods, to incorporate string digital systems and processes such as can be achieved in BIM projects. The key principles of designing accurately can lead to accurate production and off-site fits well into this process. If BIM is used on a project, but less than accurate manual methods deviate from the design, then the key benefits of using BIM are lost. MMC can enable, if used properly, a more accurate product placement conforming to the ideal BIM design. Design for Manufacture and Assembly (DfMA) techniques also fit within this principle, taking highly accurate engineering principles and applying them to the construction process.

The ethos of manufacturing is to standardize production processes and remove the risks, especially for human error, as much as possible. A good example of this is to look at the car manufacturing sector or motorbike manufacturing sector. The constituent parts used in car and motorbike manufacturing are engineered, and value engineered to their optimum. This then becomes the standard. For example, take the windscreen size of the popular Toyota Auris – the windscreen size of every one of these models is exactly the same, and, more importantly, the customer buying the car knows the size and accepts the size. If a car or motorbike customer is looking for something bespoke, they will need to go to a custom bike or car specialist and pay a lot more for the product.

Currently, many MMC manufacturers are struggling to work within the construction sector and establish this ethos. For example, a modular manufacturer could be asked to provide a large quantity of modules with a set window size. This window size requested may be close in size, within millimetres, of windows they already have in stock. Therefore, using the construction ethos rather than the manufacturing ethos, a decision could be made to use these windows instead – this is a typical construction on-site decision process. The issue may not come to light until the modules had been delivered to site and necessitated a very expensive on-site adjustment and change to all of the windows in the modules. The manufacturing ethos error is not an error made during production, it is made much earlier than that. The contract agreement should have made it clear that this is the size of window used in the modules – you can have this for X amount of cost, alternatively you can have a bespoke window that is a few mm wider or higher, but you will need to go to a bespoke manufacturer and are likely to pay a much higher price than X. The designers will make the decision to alter their design to meet the standard manufacturer's sizes to meet the cost element. Therefore, if the project designer accepts and works within the standardised sizing offered by module manufacturers, the greater will be the opportunity to realise the potential benefits.

This is an industry-wide switch that is required but, as will be discussed in later chapters, the potential for cost reduction and making the use of MMC more palatable to developers and clients, if it is accepted, it will lead to industry-wide improvements.

Another thing to discuss about MMC, and linked to standardized production, is the often-criticized lack of flexibility that comes with standardized production, and this is true, as discussed above. Every model of the Ducati Monster is the same size and shape and those in the know can recognize it as it goes past. But the industry has moved far beyond the Ford motor car motto of you can have any colour as long as its black. And while the authors are not advocating the idea of all houses looking the same, where all the design and aesthetics of buildings gets ignored, there is a place for innovation and aesthetic appeal and there is a place for standardization, sometimes called platforming, in the sector. There is potential to standardize the sizes of some elements, without the need to remove the aesthetic individuality of buildings. For example, housing developments in the UK tend to focus on just a few house types – the differences come from which house types are used where, slight changes to the external material used, and the layout of the sites. But, generally, the style and shape is the same nationwide, and this suits MMC principles.

There have been many producers of off-site products in the sector who have not survived due to the need to try and trim costs and to meet industry demands. However, if the industry was willing to flex and understand the constraints of the production a bit more and move to a more acceptable approach, then all sectors would benefit. For example, when bidding for a project a modular manufacturer suggested to the client that by adjusting the height of the room by 12mm and the width of the room by 8mm the development would meet the standard size of the modules being produced, which would make the bid X amount, which was competitive in terms of cost, and, as has been shown, was beneficial in other aspects such as speed and quality. However, the client, potentially under advice, insisted on the higher ceiling height and shorter width, taking it outside of standard production and put the cost of the module that much higher than the traditional build. Again, discussed in more detail below, this lack of knowledge from all concerned with the project is one of the reasons off-site construction is often quoted as being more expensive.

A further consideration is the potential for design for deconstruction, the principle where using off-site products, particularly modular products can allow developments to both grow and shrink as need arises easily and simply. Take, for example, an office development built using modular construction, initially built to three storeys – as the demand for offices grows in that area, additional storeys can be added (assuming the foundations were designed to take it in the first place). But then, when we have a global pandemic, such as COVID-19, and more businesses move to online processes the need for offices diminishes – rather than have a costly building sitting empty, the top floors can be removed and taken to a different location (even abroad) thereby making the product usable again. This is discussed further in Chapter 7.

The Benefits of Using Off-site Construction

There are many aspects used to assess project success and each has a level of importance that changes from client to client and project to project. A typical assessment is the use of the time-cost-quality triangle, which can be used to place emphasis on the level of importance that is most relevant to the particular project. As well as showing the most important criteria for the project, it is possible to see how each influences the other. For example, if a client and project places greatest importance on the project being finished at a specific time, then either the quality or the cost (or both) will have a negative influence on them. For example, a student accommodation build will need to be completed before the start of the academic year, to ensure students can get the rooms they want. If the start is missed then the probable outcome is that full capacity of the building will not happen for a further 12 months, losing lots of income. So, time becomes the most important aspect of this project. To ensure time is achieved, either a lower quality or a higher cost may need to be accepted, particularly if delays happen during the build process.

As well as time-cost-quality there are other indicators of project success such as health and safety impact, or environmental impact, etc. The production methods chosen for the build can all have an impact on the project success and some of the key considerations with regards to off-site construction benefits are set out below.

Time

The greater the use of off-site construction on a project, the more control can be achieved over time. For example, when timber frame wall systems are used, most of the work, such as manufacturing and detailing, is completed off-site. The only on-site activity is the erection of the components, which significantly reduces time on site. In the case of a four-bedroom house, this can cut the on-site schedule by two to three weeks. On larger projects, the time savings increase accordingly.

Panelised systems typically replace structural wall elements, but when expanded to include more building components – such as in full modular construction – the time savings become even more significant. Modular systems can deliver not just structure, but also internal finishes, services, and fixtures, further compressing the site programme.

However, off-site construction introduces different time considerations that must be managed. These are sometimes seen as disadvantages, but with proper planning, they can actually become advantages. The key issue is timing of decisions. Off-site components must be ordered earlier than traditional materials because they need to be scheduled into factory production runs. For example, walls must be fully designed and signed off well before the point they would normally be needed on site.

If a project demands a fast on-site programme but does not commit early enough to off-site methods, delays can occur. Similarly, late design changes become difficult and costly once manufacturing has started. This is often perceived as a disadvantage. However, if the client understands and accepts the fixed nature of off-site production, it can actually be beneficial. Once production begins, the design is locked (design freeze – see later in the book), which protects both time and cost from late-stage alterations.

Quality

The construction sector has long struggled with low productivity and inconsistent quality. Defects, rework, and latent issues are often expected and, unfortunately, widely accepted as part of the process. Many of these quality issues stem from the nature of on-site production, where numerous variables introduce risk and inconsistency.

Human error is one of the leading causes of poor quality on site. Workers may be unfamiliar with the site layout or specific company procedures, may lack the correct tools or materials at the right time, or may be fatigued from long hours or overtime, often added to recover delays. External factors such as adverse weather conditions or working away from home can further affect performance and attention to detail.

Material and equipment quality can also suffer. Materials may be stored improperly on site, damaged during transit, or mishandled during installation. Plant and machinery might be unsuitable for the specific task or not set up correctly, increasing the risk of poor workmanship or safety incidents.

In traditional construction, many of these problems are addressed reactively, by combining the skills of tradespeople with the problem-solving ability of site managers to work around unexpected issues. While this can keep a project moving, it also introduces variability and cost.

Off-site construction helps mitigate many of these risks. By shifting production to a controlled factory environment, processes are standardised, quality checks are built in, and external variables like weather and on-site logistics are removed. This leads to more consistent outcomes, fewer defects, and less reliance on reactive problem-solving during delivery. Also, on-site checks for quality can only be carried out sporadically due to the size of the site and the limited time available to carry out inspections, this can also cause re-work latent defects. Latent defects are defects that only come to light once the project has finished and the workforce has left the site. Typically, these defects come to light while the building owner is using the building and there is then an expensive process of project workers, often sub-contractors, returning to site and disrupting the use of the building while they carry out repairs. It is rare for the issue of the latent defect to be the quality of the material used – it is often a human error that has happened during production. So, for example, a lighting issue is rarely the light fitting itself, but more to do with the installation of the light fitting. The reason for this is the manufactured materials typically have much more stringent quality control than on site can achieve.

Off-site construction has the added advantage of these strict quality controls and these can be applied to large parts of the construction project, thereby gaining a greater benefit overall. As the products are manufactured the issues of human error can be reduced considerably. For example, if a production line is behind schedule, it is possible to increase productivity in a number of ways that does not require workers doing overtime and therefore becoming tired. In the majority of situations, the factory location is fixed and therefore the workforce is also local and not likely to be staying away from home, with those distractions, work life balance is more achievable in manufacturing than it is in on site construction. The materials used are stored in a safe specific location within the production facility, the production facility will have solid flat and clean surface for moving people and materials around thereby eliminating the potential issue of damage during production. Also, the materials used are often standardized and therefore if one is found to be damaged it should be possible to easily and quickly replace it, not always possible on site. There is also the issue of increased supervision and quality control that is possible in a factory environment than on-site as discussed above, and this can often lead to greater quality.

There is potentially an adverse effect on quality during the transportation process of getting the finished product to site and installing it. However, as these potential issues are designed into the manufacturing process it is rarely an problem. Therefore, it can be seen that both time and quality can benefit greatly through the use of off-site construction.

One of the issues relating to quality is the, until recently, lack of insurance and product assurance that is typical in traditional building process – for example, the NHBC quality standard. Recently, both NHBC and BOPAS schemes have stated they recognise the use of off-site techniques and part of their assurance schemes, which gives reassurance to developers and clients, and, in particular, those funding bodies related to MMC projects.

Cost

The cost benefit is harder to quantify and justify than the other important and key aspects of project success – a whole chapter is dedicated to this in Chapter 7. As a summary, however, it can be seen that cost can be an advantage in many construction projects if the right set of decisions and project parameters are used. For example, if a standardized sizing element is incorporated into the overall project design process, then manufacturing facilities can be more competitive, whereas if a production needs to be bespoke then the cost increases, as discussed above. Also, if on-site costs are calculated on a like for like comparison, then a stronger financial case for off-site can be made. For example, if an on-site project takes into account the financial impact of weather, which is significantly reduced when using off-site, then the case for off-site is stronger. Another consideration is the link between time and income – if a developer links the two then the significance of cost benefit becomes more apparent. Again, this is not always carried out and so the benefit is not always apparent.

Cost remains an anomaly of the on-site and off-site debate and this is addressed more in later chapters. One of the key terms being used in projects recently is that of pre-manufactured value (PMV), which provides a financial assessment of the amount of pre-manufactured (or off-site) products used in a project. There is a complexity in this, as discussed above, as it is quite difficult to categorise what is classed as pre-manufactured and what is just building materials, hence the introduction of the categories listed earlier in this chapter.

Other Benefits

The complexity of the cost issue is evidenced by several authors listing it as a benefit and others as a barrier, whereas time and quality is a clearer issue. But some of the other benefits shown from the use of off-site construction can include:

Greater Control

Part of this is the greater control that is possible in the factory environment that is discussed above, but the client has greater levels of control through being involved in an off-site project. As the project design has to be completed at an early stage, and client changes are not possible once production has started, clients have to get heavily involved in the design process and make sure they are aware of the impact of all the design decisions. This is a process shift from any projects they may have been involved with before but can give them a greater sense of control over the entire process, and therefore greater levels of ownership over the design. As it is difficult for some clients to visualize end products from two-dimensional drawings, it may be necessary to provide three-dimensional information, and if using a BIM driven ethos this can easily be achieved using virtual reality (VR) or augmented reality (AR) systems.

Long Term Partnership Development

Only a limited number of manufacturers currently provide off-site production and the number of supplier/subcontractors used on a project is reduced. Using one supplier/supply chain for the bulk of the production has the drawback of limiting competitive advantage, but the benefit of creating a strong working relationship. As more projects develop with the same partner, the speed of working together and the insight into each other's needs becomes clearer. This will lead to efficiencies overall, in both time and cost elements.

Increased Worker Safety

One of the most important considerations on any construction project is worker safety. The industry remains one of the most hazardous sectors to work in, with high levels of risk associated with manual handling, working at height, and operating in dynamic and unpredictable environments. Modern Methods of Construction (MMC) provides a significant opportunity to reduce these risks by transferring a larger proportion of activity from traditional sites into controlled factory settings. In this environment, repetitive tasks can be carried out using mechanisation, automation, and standardised processes, which reduces exposure to accidents and improves overall working conditions. The benefits are not only human, in terms of fewer injuries and improved wellbeing, but also financial, through reduced downtime, fewer claims, and more predictable programme delivery. This dual impact highlights how safety is both a moral and commercial driver for innovation.

Safety is also increasingly embedded within the UK's legislative framework. In 2022, the UK Government introduced the Building Safety Act, which was designed in response to high-profile failures within the built environment. The Act strengthens accountability by placing new duties on designers, contractors, and building owners, ensuring that safety is considered throughout the lifecycle of a building. Central to the legislation is the concept of the golden thread of information. This requires that accurate, up-to-date, and accessible data about a building's design, construction, and maintenance is created and maintained digitally. MMC, and in particular when linked to the use of BIM on projects, can play a key role in supporting the golden thread because factory-based production works with detailed digital models, component specifications, and quality assurance records as part of the manufacturing process. These outputs can be integrated directly into the project plans, creating a traceable and consistent flow of data from design through to installation and maintenance. By embedding digital records at each stage, MMC helps ensure that the golden thread is not just a compliance exercise, but a practical tool for managing safety and performance over the lifetime of a building.

Taken together, the health and safety advantages of MMC and the regulatory requirements of the Building Safety Act demonstrate how innovation and legislation are working in parallel to raise safety standards across the industry.

Reduction of Ecological Impact

There is a potential for reducing the impact of construction projects on the natural environment by using off-site techniques. Interaction with the natural environment is likely to cause some disruption. For example, this could be spillages, run off, or even just the noise and activity on site disrupting local wildlife. Reducing the on-site work limits many of these aspects and this will also be discussed more in later chapters.

Improved Waste Reduction

Standardizing the build process will reduce the amount of waste that is produced – this is because standard sizes can be developed to match the material production facility. For example, making the walls match the same height as the plasterboard will reduce the off cuts in plasterboard, etc. Off-site techniques will make this easier to do if utilized correctly. Also, when manufacturing in a factory environment it is much easier to arrange, store, and reuse any off cuts that do happen. On site production typically will lead to off cuts getting damaged and eventually skipped – this will lead to additional cost not only in terms of finances but also in ecological costs when the waste goes to landfill. Reducing this is a major benefit. The cost aspects are addressed in greater detail in later chapters.

Climate Change

The weather will be shown later in this book as a key cost impact on construction sites, as it can reduce the productivity of workers. Climate change is an all-encompassing term used to discuss the long-term expected changes in weather patterns, with the prediction that excessively hot days will increase and there will be more storms, heavy rain, and floods. Therefore, any impact seen on sites today is likely to get worse in the future and changes will need to be made to cope with this. Off-site manufacturing is one solution, for two main reasons: the work is carried out inside and therefore is not overly affected by weather patterns, but also the production is seen as more efficient and therefore has less impact on climate change.

Resource Depletion

Using off-site construction allows for greater consideration in using sustainable sources of materials than perhaps can be considered on site. The more controlled environment allows for greater levels of innovation so that more consideration of the environment and sustainable resources can be taken into account. Also, off-site construction can provide greater levels of insulation without impacting overall floor space reduction and therefore it is possible to have more energy efficient homes and buildings, thereby saving on resources used to provide electricity and gas. There is also a reduced need for water when manufacturing compared to on-site production, even more so if good water harvesting techniques are used in the factory, water being another valuable resource to be saved

Aging Population

The fact that people are living longer now than in the past means that we have an aging population, and this leads to different needs. In the past a smaller footprint home with stairs was desired, but now there is a greater need for single storey accommodation to aid those with the reduced mobility issues that can come with age. And while, in and of itself, this does not show the benefits of off-site construction, using modular systems can help with adjustments to properties as the needs of that property change. So, for example, a small two-storey house may be suitable for a young couple, who may then add some rooms easily (with a modular system) as they start a family, and then as the children leave home the building can be reduced and moved to a single storey as needed. Also, in contemporary societies there is no longer a set 2.4 children family type – society has changing needs, with more people willing to share accommodation, and having different requirements such as no garden, or no parking spaces. Modular production, if utilized correctly, can help to adjust accommodation to meet these changing needs.

Lack of Skilled Workforce

Finally, the benefit that is often presented is that of the reduced need for a highly-skilled workforce. As processes in a factory environment can be standardized and training can be on more specific aspects than is often possible on a construction site, then there is reduced need for the highly-skilled, on-site workforce that is currently deployed and in short supply. The skills gap that is currently being experienced on construction sites has two key impacts: the first is that wages are rising, as skills become more in demand, basic economics lead to a greater cost associated with this. The second impact is in rushing the production – on-site construction is known for the potential for low quality production, and part of this is because the on-site labour is trying to do more in less time. Some of this time reduction can be countered through new production processes, but much of it is the product of rushing projects to meet demands with a reduced skilled workforce, and worrying about snagging at a later date.

Chapter Summary

Modern Methods of Construction, sometimes called off-site construction, Design for Manufacture and Assembly (DfMA) amongst many others, is the process of designing and producing components for a building project in a factory environment. There are many benefits to doing this, if the process is used correctly, and also if the on-site aspects of the project are produced in the right way to correctly incorporate the off-site elements. The use of digitalization, including processes such as Building Information Modelling, AI enabled quality insurance systems such as sensor tracking, etc. (see Chapter 5 for more detail) has enabled the correct use of off-site products more effectively.

New legislation that ensures the safe design, development, and installation of components also lends itself to the use of MMC. But it is important to note that the systems must be designed properly and in very close collaboration with the overall project design team. The design team must ensure they design the project to best utilize off-site components. The installation team must develop awareness and understanding of the systems being used, and a greater awareness of the increased accuracy required to work effectively with off-site systems.

The government developed a list of seven categories to ensure greater compliance with utilizing MMC into projects and each has their benefits and drawbacks – for example, modular systems including pod systems have more of their elements manufactured in a factory setting and therefore should be of higher quality (see Chapter 6). The use of panelised systems and components is already well-established in the industry and so the greater knowledge of on-site staff is already developing. Later chapters will explore how each of these categories will impact on procurement (Chapter 3), time (Chapter 5), quality (Chapter 6), cost (Chapter 7) and supply chain management (Chapter 4).

References

Building Safety Act (2022). www.legislation.gov.uk/ukpga/2022/30?timeline=false&view=plain (accessed August 13 2025).

Cabinet Office (2020). *The Construction Playbook: Government Guidance on Sourcing and Contracting Public Works Projects and Programmes*. Cabinet Office.

Excellence, C. (2022). *Value Tool Kit*. https://constructingexcellence.org.uk/value-toolkit/#:~:text=The%20Value%20Toolkit%20enables%20value-based%20decision%20making%20focused,for%20Value%20report%20led%20by%20Ann%20Bentley%20CBE (accessed August 25 2025).

MHCLG (2019). *Modern Methods of Construction: Introducing the MMC Definition Framework*. London: MHCLG.

Myers, D. (2017). *Construction Economics: A New Approach*. New York: Routledge.

Chapter 3

Procurement

Scope of Chapter

Procurement is a key aspect of any construction project, it is the process of acquiring the goods in order to complete the project, but it is also the process the client goes through in order to acquire the overall project itself. There are many things that need to be considered when deciding on a procurement route for the project amongst them are the requirements in time, cost and quality that the client places on the project, and as these can be affected by the use of MMC, it is an important aspect of this book and needs consideration. The main procurement routes of Design-Bid-Build, often termed traditional procurement, is a key aspect in the industry and is still one of the most used procurement routes. This type of procurement allows that all design should be completed before contractor involvement and before build starts. Design and Build is another commonly used route and allows for very early contractor involvement and the design developing as the build progresses. A form of procurement that allows for a construction specialist to be appointed either as an advisor to the client or as a project manager to the project are termed management procurement routes. There are many other types of procurement, each with their own perceived strengths and weaknesses for construction projects, and for MMC projects in particular, and again this is addressed below.

It is important to note the particular limitations that using MMC can place on a project, and how these can be either countered or exacerbated through the choice of procurement route. One of the most significant factors is the need to order components early in the programme, as the manufacturing process requires long lead times compared to traditional on-site methods. This in turn means that projects using MMC often require an early design freeze, limiting the opportunity for later design changes. While this can reduce uncertainty and improve delivery, it also demands closer coordination between the client, designers, and suppliers at the outset.

MMC frequently involves adjustments to contract and payment terms. For example, payments may need to be structured to reflect the manufacturing and delivery schedule of off-site components, rather than the more traditional staged payments linked to site progress. Early supplier engagement is therefore critical to align expectations, clarify responsibilities, and avoid contractual disputes.

These issues have been recognised at government level. The UK Construction Playbook (Cabinet Office, 2020) sets out policies to encourage the adoption of MMC through procurement, highlighting the need for collaborative approaches, clear pipelines of work, and evaluation processes that reflect whole-life value rather than lowest cost. The emphasis placed on procurement in national guidance underlines why this topic is so timely, and why it plays such a critical role in the successful delivery of MMC projects.

DOI: 10.1201/9781003570431-3

Project Life Cycle

In order to appreciate the impact of production type on procurement method choice it is necessary to understand a typical project cycle. Figure 3.1 shows a traditional project cycle, based on a Design-Bid-Build type of procurement and while the process and cycle will change slightly depending on project type, it is easier to start with this version. Nearly all projects start with an idea, typically an idea that relates around satisfying a need. For example, this could be a housing development, or a factory extension, or a hospital wing, a new bypass, or new reservoir – all are trying to furnish a need that will typically fall into either a civil/society need, an economic need or environmental. These are often categorized as the three pillars of sustainability.

Inception

Once the idea has taken hold the client, sometimes termed employer, would engage with a design team, typically the designer such as an Architect, Architectural Technologist, Civil Engineering or Architectural Engineer – depending in the project type. During this initial meeting the general scope of the project will be expressed, such as the type of project, perhaps the scale of the project and location, and also budget will be addressed at this point. Another key aspect during this meeting is an initial exploration of risk, and if there are any specific requirements, such as high quality, or specific aesthetics or utility, etc.

Concept Design and Brief Development

Once an agreement to proceed with the project following the development of a basic inception knowledge is achieved, often the design team will start to create the basic details and sketches, and the feasibility study will take place. The level of work carried out in this section is very important for ensuring project success and some major decisions are finalised, but the quantity

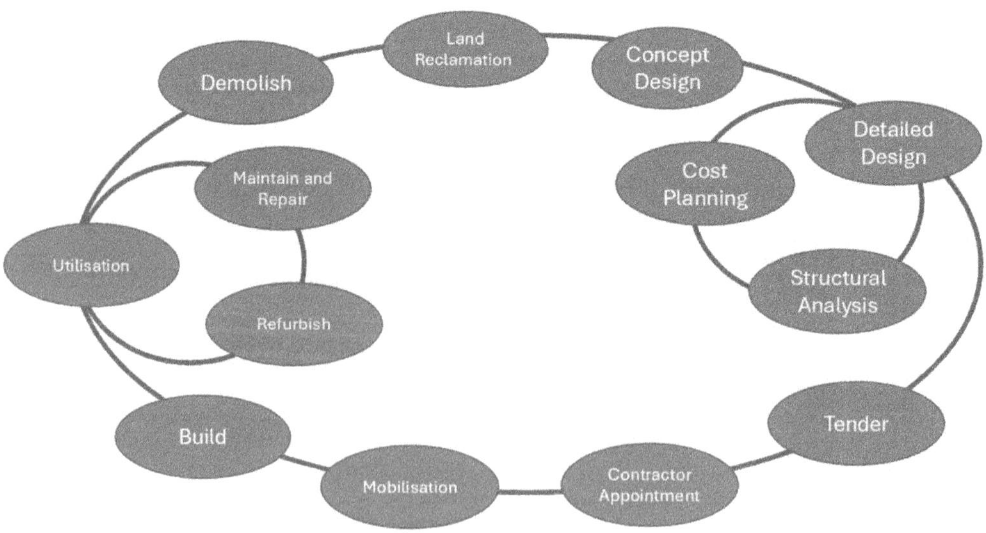

Figure 3.1 Project Life Cycle

of work is still reasonably low level which would allow for clients to still make a decision not to proceed without having incurred too many fees. Even if alternate procurement routes are chosen, such as Design and Build, this part of a project may still be undertaken with a design team support. With regards to the use of off-site techniques, the early stages of a project are often the best time to decide on their level of use, as future design choices will be shaped by what can be achieved within the confines of the manufacturing process. In this wider context, procurement has a central role in linking construction to manufacturing principles, particularly through the use of standardisation and pipeline volume. By leveraging repeatable design solutions, buyers can create opportunities for manufacturers to invest in more efficient production processes, ultimately improving the commercial viability of MMC.

An important emerging idea in this area is the "platform approach" to construction, where clients procure a set of standardised components or modules that can be applied across multiple projects. This approach, which relates directly to the MMC categories and the concept of repeatability, encourages greater consistency and scalability in delivery. When procurement is co-ordinated across programmes or even between departments, it creates a steady pipeline of demand. This, in turn, allows factories to optimise production, achieve economies of scale, and reduce unit costs, while also enabling quicker turnaround times and more predictable outcomes in terms of cost and schedule.

Recent government initiatives have highlighted the potential of this approach. For example, the Infrastructure and Projects Authority (IPA, 2021) has advocated for platform-based solutions as part of its push to modernise construction and pan-government procurement frameworks are being explored to support the adoption of MMC components at scale. Such initiatives demonstrate that moving away from one-off, isolated projects and adopting a strategic, collaborative approach to procurement can deliver more efficient manufacturing processes and better outcomes. By thinking beyond the boundaries of a single project, clients can help align construction with manufacturing practices and strengthen the link between procurement choices and successful delivery.

The brief is one of the most important documents and is developed at this stage – it sets out the key parameters for the development of the project such as size, quality requirements, technology requirements, budget, key performance indicators, utility requirements, etc. The brief needs to be flexible enough to allow the client to make informed decisions on future development direction, but rigid enough to direct the design and production aspects of the project towards an efficient outcome.

Also, at this stage it is common to put a bit more detail into the cost of the project – this can be done in a number of ways including the use of functional or superficial unit estimating. These early project cost techniques provide a rough guide to the overall cost of the project but are based on very little information. The functional unit of a property is a set number requirement based on the function – for example, number of car park spaces, number of students in a school, etc. The superficial unit method uses M^2 floor area as the basis for costing. There is very little that can be done by cost managers or quantity surveyors to allow for size, shape, layout or quality requirements in adjusting these as at this stage, as these are not known. As such the figure provided is very vague, this is made even more so when using off-site production as the level of cost data available is much harder to come by and, as will be shown later in this book, what is available does not appear to be accurate. One of the issues relating to this vague cost, is that often they are used to set the budget for the entire project, and once a budget is set it will guide the decision making for the rest of the project too. Therefore, the lack of accuracy places restrictions on the future project. It is also at this stage that, typically, a procurement route is decided upon (see below for more discussion).

Detailed Design Phase

This phase is where all of the technical, utility and aesthetic aspects of the design are developed – it can take a long time to get through this stage and will depend on who is controlling the process and what type of procurement route is decided upon. There are several aspects that are incorporated within this phase, but the key concept is to take the project from an idea to a stage where it can be built. It is common in contemporary construction to use digitized methods, and these will often follow the process as set out in BIM protocols.

Design

This is the detailed visual aspect of the building project, setting out what the project size is and what its orientation is. The key driver for this is client need and utility – for example, if the client needs a new school that has classrooms able to cater for 40 children at a time, then this will govern the size of the room, and, ultimately, the building.

Technical Needs

This is where technology requirements will govern size – for example, the standard size of windows or furniture, etc. – this is a particularly important aspect to consider if using off-site construction. As off-site construction becomes far more effective as a choice if standardized sizes and layouts are incorporated, this may govern the overall design of the project.

Energy Use

This is linked to orientation as well – the environmental performance of the building can be greatly affected by the design, and in particular the orientation of the building. There is greater scope for the use of energy efficiency measures such as energy efficient heating and power systems, including renewable energy, better insulation, etc., but also putting the parts of the building that need light and heat from the sun – to be facing south can also make buildings perform better.

Legislation

There are legal requirements that have to be met – the construction industry is one of the most highly regulated and there are many aspects that relate to the design, such as size of windows, amount of ventilation, fire management, etc. The building regulations, environmental regulations, health and safety regulations, and planning regulations have the biggest impact, but all aspects need to be complied with.

Structural Design/Detail

Once the design is partially complete, the drawings will be passed to the structural engineer who will carry out a series of structural calculations to ensure the finished building product is safe and secure and will withstand all of the forces and loads placed up on it as part of the building use. These forces and loads are numerous and include dynamic, live and dead loads, snow and wind loads, thermal loads, etc. The calculation process is complex and, as the building product and design changes, the calculations will need to be re-processed time and time again. An additional consideration is that, once the calculations are complete, this may change the original design – for example, if the foundation type needed to change, or if layouts need to be altered – and this can lead to several iterations of intervention. This is made more complicated with the intervention of the cost manager, structural engineer, and modular production and design teams.

Cost Planning

This textbook is going to focus on many of the aspects that happen during the cost planning and cost management processes, but to provide some context for the rest of this chapter and book it is a good idea to introduce it now. The cost planning process is carried out during the design phase of a project to ensure the design team are keeping within budget, and, if this is not happening, inform the client of this, and provide advice on alternative solutions to bring the design back within budget – this will allow the client to make informed decisions on whether to stretch the budget or change the design. As stated above, this change to design can potentially then have a knock-on effect on the structural calculations and vice versa. As such it is acknowledged within the sector that the lack of collaboration between the various design team factions can lead to inefficiencies in the design. The advent of BIM processes has led to greater and quicker levels of collaboration so that the whole process is more efficient and should lead to improved outcomes and project success. But more on BIM later.

Out to Tender

Again, looking at this from a purely traditional perspective, the contract documents that have been prepared in the process above would be collated and sent out to the contractors for pricing depending upon the procurement type that is used. The contract documents in a traditional contract would typically include:

Plans and Drawings

These are a visual representation of the project and will provide detail such as size scale and building element association with others. Traditionally these would be produced in a two-dimensional format and would typically have been paper based a few years ago. Nowadays, with the better digital infrastructure available pretty much globally, the preferred method is to use and distribute digital drawings. However, for those entering the industry it is important to note that there are still projects, particularly smaller projects, that still rely upon paper-based drawings. While the transition is still happening from paper based to digital drawings it is important for new entrants to develop skills in reading and understanding both forms of media. To add complexity there is now a growing element of three-dimensional drawings and systems available for use digitally. This is developing alongside the growing use of BIM systems and process. While government-led projects in England have been required to use BIM it is still not growing as fast as expected in the industry.

Specification or Schedule of Work

If the drawings are the visual representation, then the specification is the written detail on each of the elements of work to be carried out. This will explain the materials to be used and any specific installation aspects that need to be considered. The detail on the materials will not only set out the colours and types of material, but also express any key regulations or quality standards that need to be met by the material. This will help if alternative materials need to be sought later in the project, if, for example, the material is no longer available, particularly an issue in long term projects that have a long lead in from the specification development to the material purchase. As modular and panelised systems are factory produced, storage becomes less of an issue and can help to reduce the risk.

The specification can be in a written document, or it may be provided within the drawings (on smaller projects). However, with certain projects it can be provided as a schedule of work which is actually a pricing document.

Contract Documents

An often-forgotten aspect of the tender documents are the contract documents themselves – these are however key aspect of the tender. For one thing they will express key costing information, such as

- *Time scale*: The time that the contract is to be started and completed in – as has been shown earlier, the link between time and cost is well established and the quicker something needs to be built, often the more it will cost (more on this later).
- Liquidated and ascertained damages (LADs): These are financial penalties that are imposed on a contracting party if they fail to complete the project by an ascertained time. The more important the completion time is to the client, the higher this penalty is likely to be. But we must also consider that the higher the LAD penalty is, the higher the price is likely to be submitted during tender – as the contractor is carrying greater risk, they either have to increase costs to make sure they will not miss the deadline, or take out an insurance should it happen.
- *Special working conditions*: These can include things such as deliveries only happening at a certain time of the day, or working hours being restricted or special consideration of pollution control, etc. – all of which can increase the cost of the project.

Documents for Pricing

These can take many forms and again will depend on the type of procurement however for ease they are explained in this section.

Bills of Quantities

In years gone by this was the most used format of pricing document. It is where a quantity surveyor would measure elements of work from drawings and provide this in a Bill format. The focus is on elements of work not materials as the costing of these elements also requires labour and plant costs to be included. So, for example, it may say $100M^2$ of brick wall as opposed to 6,000 bricks x amount of sand, etc. There are, or were, issues with this type of costing document in that if errors are made then often it is the client that picks up the cost, and the quantity surveyor who lost their reputation. Also, it is an expensive process as the quantity surveyor can spend many months on producing the bill ready to go out to tender. Due to these risks and costs, it has become a more common practice now to request an alternate to the Bill of Quantities to be produced by the contractor – this is often called a Builder Quantities Format, and this removes both the risk and expense of items being missed from the client, as the contractor produces the document and then prices from it. However, there is a risk involved in this process as well – for example, any assumptions that are made during the production are specific to that contractor (as opposed to in the traditional way every contractor works off the same assumptions) and this can lead to a wide variety of tender responses and a longer, more complex and more costly tender analysis process. As such, Bills of Quantities are still used in some projects.

Schedule of Works

This is part specification and part quantity listing – again it is based on elements of work, and is most used in projects that have a series of small elements, such as a maintenance, or repair/refurbishment contract. The quantity of the element is mixed with the detail of the work that is needed listed and priced.

Schedule of Rates

In maintenance type contracts, where the quantity of work is not known prior to the contract start, a list of possible elements of work is often provided and priced. A good example of this would be where a contract is offered to carry out all of the repair work for a social housing provider. The employer/client would know that there will be a need for a set type of work, door handle replacement, light fitting replacement, kitchen work top replacement, etc., in all of their properties, but at the time of going to tender it is not known how many of each. Therefore, the tendering contractor would price up based on doing one aspect. But then as the work comes in, the work is done and the contractor gets paid for doing the work. One of the issues with this type of contract is that the client cannot benefit from bulk purchase savings, etc. Note there are published books with the type of elements of work that may be needed, and the contractor can use these to provide their cost elements.

Contractor Appointment: Contractor Selection

This is the start of the detail relating to procurement and tender selection. The process of appointing the contractor will depend on many variables including but not limited to:

Early Engagement of Contractor (ECI)

Appointing a contractor earlier in the project timelines, than may happen traditionally has been shown to have some benefits. There are also some drawbacks in that it adds an additional line of communication, which can add cost and slow down the decision-making processes. There is also the disadvantage of having to appoint a contractor before clear design decisions are made which is risky.

Early contractor involvement can, however, offer several advantages to construction projects by allowing the contractor to contribute expertise during the early design and planning stages. One of the main benefits is improved buildability, as potential design and construction challenges can be identified and resolved before work begins on site. This reduces the risk of errors, redesign, and costly delays later in the project. ECI also strengthens risk management by enabling risks to be allocated more appropriately between the client, designers, and contractor, which helps to minimise disputes and improve certainty of delivery. From a cost perspective, contractors are able to provide accurate market-based pricing and support value engineering exercises at an early stage, which can deliver both cost certainty and long-term value. Programme efficiency is another advantage, as early planning of sequencing, logistics, and procurement allows activities to be streamlined, particularly when Modern Methods of Construction (MMC) are being used and manufacturing slots must be secured in advance. In addition, ECI encourages innovation by bringing specialist supply chain partners into the process earlier, facilitating the integration of digital tools such as BIM and supporting sustainability goals. Importantly, the collaborative nature of ECI can also reduce conflict between stakeholders, replacing the adversarial relationships often associated with traditional procurement with a more cooperative approach. Overall, ECI provides clients with greater certainty, reduced risk, and stronger alignment between design and delivery, making it particularly valuable in complex or innovative projects.

Client Preference/Experience

If the client has had a good or bad experience in the past with a particular procurement or tendering method, then this may influence their decision on the process to follow for the future. It might not even be construction procurement experience that could sway the decision – for example, in the past they may have had a bad experience when passing control of a project to someone else, and therefore have become risk averse, and a need to keep control of the project close.

Advisor Preference/Experience

This is a follow on from the above – typically it is a project manager role or a quantity surveying role that advises the client on the best form of both procurement and tender process to follow, and their experience of success and failures from past projects is likely to sway their decision. Ideally each project is taken on its own merit, but it is difficult for humans to not take past experiences into account when making decisions.

Time Cost Quality Focus of the Project

This is potentially the most important and influential consideration on procurement and tender method choice. As will be shown below, each of the options has a particular slant towards improved time, cost or quality outcomes and therefore over and above the discussion above, will have a very strong influence on the final decision.

Level of Acceptable Risk

Each of the tender methods, rather than procurement routes themselves, will distribute risk differently between the client and the contractor, and this has a direct influence on the choice of tender method. The balance between transferring risk and maintaining a competitive price is always difficult to achieve. For example, under a lump sum tender the contractor accepts responsibility for both realised and potential risk. While this provides the client with greater cost certainty, the contractor will typically increase their price to include contingencies for unforeseen risks. In general, the more risk placed on the contractor, the higher the tender price, though this can also provide the client with greater budget certainty.

The amount of control a client wishes to retain also shapes this decision. Some clients are more risk-averse and want to maintain control over design and build choices. This approach is often linked to projects where quality is prioritised above either cost or time. However, increased control usually means a greater degree of client input into decision-making, which can slow productivity and potentially increase costs.

When applying these principles to Modern Methods of Construction (MMC), additional factors need to be considered. Because many components are manufactured off-site, contracts often require early payments to secure factory slots or cover material costs before delivery to site. This shifts part of the financial risk profile compared to traditional projects. To mitigate these risks, bespoke contract provisions are commonly used, such as vesting certificates, which transfer ownership of off-site materials to the client once payment is made, or off-site material insurance, which protects against loss or damage while in storage or transit. In some cases, project bank accounts can also be used to ensure that funds are safeguarded and released fairly. These mechanisms provide reassurance to both client and contractor, supporting the greater levels of upfront commitment that MMC typically demands. It is important to ensure risk is not only allocated but incorporated into contracts so that clear lines of responsibility and liability are set out from the outset, this is explored more in Chapters 4, 5 and 6 later in this book.

Once these aspects have been taken into account the decision needs to be made regarding procurement routes and tender method, below these are discussed in conjunction with each other as there is quite a lot of cross over between the two concepts.

Design-Bid-Build

This is often called traditional procurement as until recently it was the most used form of procurement and has been used for centuries. The basic premise is that all of the design and specification for the build is developed prior to going out to contractors. This has some advantages to it but also some disadvantages, bear in mind the criteria set out above.

Drawbacks

The process of getting the drawings and specification ready for sending to contractors can take a long time, particularly if it is a complex build that has implications for planning approval, regulatory approval, and makes the Design/Cost/Structures process longer and more difficult. Often the design time on a DBB project is longer than the actual build time. Another key drawback in this type of procurement is the lack of early contractor engagement, as discussed above. A further disadvantage is that it can lead to a more costly design, due to the increased length of time and the increased input from consultancy professionals.

Advantages

One of the key advantages, and a reason for its use as the number one procurement system for so long, is the greater levels of control that is provided to the client. As the design and other contract documents are complete prior to contractor involvement all decisions are based on client need, under the guidance of the design team including architect, technologists, quantity surveyors and structural engineers, etc. This will often lead to the design of the building taking into account all of the client requirements such as aesthetics and utility or use of the building, before decisions on buildability. It will also help in ensuring the right level of quality can be achieved. There is also an increased possibility of a smoother build process as all decisions have been made prior to starting and the finished product can be seen by the client, this benefit is enhanced with the advent of the use of BIM techniques and digital technologies such as virtual reality and augmented reality.

Implications and Tender Options

The decision by the client and the design team to use a DBB procurement route is a decision to increase the time of the projects, increase the cost of the design process but equally increase the quality of the finished product. As it is such a well-known and much used procurement route it does also lend itself to a variety of tender process options. A lump sum contract is a typically used tender method – this is where the contractor takes all of the cost risk onto themselves and provides a figure covering all contingencies. As the design is fully complete, the perceived risk is reduced and therefore contingencies are reduced keeping the price more manageable. Partnered contracts can work with DBB, however the impact is reduced as the scope for making joint decisions to beat a target cost is also reduced (as all decisions have already been made) but one method that can be used, and often is on smaller jobs, is the negotiated tender method. Approaching one contractor with all of the design information and negotiating a price is easier when all of the design is available.

What does this Mean for MMC Projects?

DBB procurement does work with MMC projects, but it needs to be a reasonably early decision on behalf of the client to go down this route and the potential increase in time needs to also be considered. If off-site construction techniques are to be used, then the production line timeframe needs to be considered in getting the products – this is potentially lower for timber frame products, than it is for modular products, but still needs to be considered. In a DBB contract, typically, the contractor is appointed with only two to four weeks mobilisation time (the time from when the contractor is appointed to when they start on site). During the mobilisation time the contractor gets their work force in places, appoints key sub-contractors and orders key materials. MMC can have a production line time frame of much longer than this, which means either the mobilisation time needs to be extended (thereby extending the project), or the MMC products need to be ordered before the contractor is appointed. This is possible to do using systems such as nominated supplier contracts, but it does place more risk back onto the client. Therefore, the big implication in using MMC in a DBB contract is the increase in time or the increase in risk.

Design and Build (D&B)

The Design and Build (D&B) procurement route is becoming more popular, with a recent CIOB publication stating it is now used more than DBB. The key concept of the D&B form of procurement is that the contractor gets involved in the design of the building and the design can therefore continue to be developed after the production of the building has started. For example, the foundations can be designed completely and while the internal layout of the building is finalized, or while the internal finishes are decided upon, the work on site can start. There are many different forms of D&B and can sometimes carry different names (such as Design and Construction, Novated D&B, etc.) but the key difference between each is at what stage does the design move from the control of the design team into the control of the contractor. When Design and Build (D&B) was first introduced in the 1980s, it typically did not involve a dedicated design team. Over time, it became clear that this limited the types of buildings it was suitable for, mainly to those with little or no aesthetic requirements, since contractors often prioritised ease of construction, resulting in basic, box-like designs. The recent developments in D&B however mean that the design can now be started by the design team and under the client's control (similar to DBB above) but then at an agreed point in time the control of the design moves over to the control of the contractor – often this is done by novating the control of the actual design team that has been working on the project and putting them under the control of the successful contractor. Novating is a term used to denote moving a bi-party contract, so that one of the parties is removed from the contract terms but a different party takes their place (here this is removing the client from the contract terms and responsibilities and replacing with the contractor).

Disadvantages

One of the key disadvantages with D&B is the lack of control the client has over the design process, which is probably the reason for the development of the new systems discussed above. For a contractor, getting involved with a Design and Build contract carries greater risk (as they are taking on more responsibility, this typically comes with more risk), and therefore they will often look for ways, when they have control of the design to make more profit, often this can lead to changes to the client's needs. To counter this happening, the client will often employ a consultant through the project duration to look after their interests and make sure the building ends up in the situation they require, and this will add costs to the project. Often this consultant is termed an employer's agent and will often be a professional with financial management knowledge typically a quantity surveyor, or sometimes a project manager from a quantity surveying background.

As well as the disadvantage shown above in potential quality impact, D&B also has the disadvantage of not have as much cost certainty. There is no direct evidence on whether the cost of the project is higher or lower using this form of procurement, but the cost certainty is reduced. With DBB the design is complete before getting tenders, all parties know exactly what is going to be produced and how much it is going to cost, prior to entering the contract. But with D&B some or all of the design is not completed prior to entering the contract and so there is no way of knowing how much the finished product will be.

Advantages

The biggest key advantage to this type of procurement is the reduction in time for producing the overall project, as the design and the build can continue alongside each other – the time for each aspect of the production can overlap reducing time of build. Depending on when the design team is novated into the project, there is also the added advantage of getting early contractor involvement (ECI) in the design – this makes the design take more note of production processes and so the production process itself is likely to be smoother. There are other advantages of ECI such as greater knowledge and contacts with a wider sub-contractor base. The sub-contractors are often the organizations that carry out the work, and ECI can often lead to early sub-contractor involvement too and so a more specialist level of knowledge can also be included in the design phase.

Implications and Tender Options

The reduction in time can be a major advantage depending on funding sources, in some publicly funded projects, there is a finite time in which the money must be spent and using a procurement route that allows for speed of build becomes a more important advantage. The stage at which the responsibility moves from the design team to the contractor will have an impact on several aspects. For example, the contractor organizations are often keen to do Design and Build projects as having control of the design can lead to better ways of building projects, leading to a more profitable outcome. However, the more that the design team retain control the less the contractor can take advantage of these opportunities, and the bigger their price will become and the fewer organizations will be interested in the project. Having control from the start, they can govern the shape of the building, which will lead to an easier and more profitable outcome (it is easier to build a square box that a round building). But having the control of design over at Stage 4, where it is just fixtures and fitting to finish, will not leave the contractor a great deal of scope – in this case less contractors will be willing to tender, and those that do will treat it very much like a DBB and their prices will reflect the higher price inherent in taking on the responsibility of design, but not any of the advantages. The tender options for this can also include competitive bidding, negotiating may be an option and it is common to have this as part of the framework agreement/partnering process. In a framework agreement the relationship between the client and contractor is well established and so the trust and respect needed for D&B to work effectively is already established.

What does this Mean for MMC Projects?

The key advantage of D&B is the speed in which the building can be created and the reduction in time that is available for the overall project. Adding off-site production into a project can help with reducing time on site, but it is important to allow for the additional time needed to order the products and get them delivered to site – production line time (PLT) needs to be considered. Depending on when the design is novated to the contractor (which is when the contractor is appointed) will provide the level of impact PLT will have on the project time. For example, if the contractor is involved in the design early on and wins the contract at Stage 1 or Stage 2, then they have early warning of the need to order the off-site materials and so any PLT impact is reduced, or even negated. The contractor knows very early on that off-site is required, can place the order early and continue with the rest of the design, mobilization and early part of the build (often ground work and sub structure) allowing for the production of the off-site components to continue in this period and hopefully not lead to a delay waiting for delivery. However, as discussed in DBB above, the closer to mobilization that it is before the contractor is appointed the longer the delay will be (unless the client places the order earlier and nominates the supplier in the contract). In summary D&B works very well for MMC production, as long as the contractor is appointed early enough in the project and the decision to use MMC is also early.

Design, Build, Finance Operate (DBFO)

This is just one type of a suite of procurement options that is often grouped into an overarching term of Private Finance Initiative (PFI). The basic principle is that private finance is used to fund public projects – private industry put up the initial funds to produce the project and then recoup the costs plus the profit from the life of the project. This can be either by leasing the project back to the government over a set period of time (this might, for example, be used for building a new school or hospital), or it might be by charging for the use of the project for a set period (building a new road (toll road) bridge (Severn Crossing for example) or tunnel (Channel Tunnel, etc.).

The system has been much maligned in recent years due to the size of the returns being reported from private organizations for getting involved with these projects, thereby costing the public more than expected. Speed is often a big consideration in these projects as the quicker they are finished the quicker the funding body starts to reclaim their money – as such, off-site techniques, if considered early enough, would be a good option for these types of projects. Although it has to be noted that due to poor reviews it is not used as often as it may have been done in the past.

Emerging Cost (Maintenance Contract)

This type of procurement route is used mainly on management contracts, at the start of the project and as the work is awarded through the tender process (often using open tendering) – the amount of work to be done is not known, and the actual work content itself is not known. The contract works by pricing up standard work that may occur during the length of the project and this is the basis on which the contract is awarded (for costs, other criteria are often used as the key judgement in this type of contract). Then, as the project commences and the items of maintenance appear, the contractor is paid for the work they do based on the prices submitted. For example, a city council seeks a contractor to run the repairs and maintenance on all of the council owned properties – once it starts, the requests for repairs (door handle broken, windowpane smashed, etc.) are sent to the contractor, completed and then invoiced.

This type of contract has no ability to forward plan the work in most of its cases as it is reactionary in nature – as such it is not really relevant to a system that requires early purchasing of materials or products (as is the case with MMC systems)

Framework Agreement

This type on the other hand can be very relevant to MMC types of contracts. A Framework Agreement is similar to a tender list in that it appoints a group of contractors for a set period of time who will then be included in the procurement of future projects of that type. Typically used on larger projects such as a schools framework agreement, or highway maintenance framework agreements, it does allow for the long lead in to the projects – scrutiny of the contractor happens in one go, when adding to the framework, as such the qualitative aspects of procurement is shortened and focus can be made to the cost aspects, often this can also lead to target costing as the tender procedure. Either way, the ability to utilise the benefits of off-site construction can be seen. It is also noted that often government-based projects are developed with the owner being the end user, and, as such, quality of finish is an important consideration. For example, the government holding a schools framework agreement will be the organization that operates the school running costs, and therefore these are important and can be an important consideration moving decisions towards the use of MMC. But in a housing development, the developer sells the house on to an end user, and so for the developer the running costs are a secondary consideration to the financial returns and marketability of the house.

Final Thoughts on Procurement Routes

A further critical consideration when procuring Modern Methods of Construction (MMC) is ensuring that the quality of products and systems is addressed from the outset. Unlike traditional construction, where much of the quality assurance process can be observed and controlled directly on site, MMC relies heavily on factory-based production. This shifts the focus of quality management upstream, into the supply chain, and places greater emphasis on procurement decisions. In this respect, procurement is not simply about cost or programme, it is about selecting suppliers with the right capability, accreditation, and quality management processes to deliver reliable and compliant products.

One of the most effective ways to achieve this is through supplier pre-qualification, where tendering organisations are required to demonstrate adherence to recognised standards before being considered for contract awards. Within the UK, schemes such as Build Offsite Property Assurance Scheme (BOPAS) or NHBC Accepts are increasingly important. These accreditations provide confidence that the products, systems, and processes used by a manufacturer have been independently assessed, tested, and certified as fit for purpose. Including such requirements as part of the procurement criteria ensures that only suppliers who meet robust, industry-recognised standards are eligible to bid. For housing projects in particular, this provides additional reassurance to clients, regulators, and financial institutions that the systems being used will be durable, safe, and insurable over the long term.

Beyond pre-qualification, procurement can also directly mandate quality assurance practices that extend into the manufacturing environment itself. For instance, factory audits and detailed quality plans can be required as part of contractual obligations, ensuring that suppliers adopt rigorous inspection, testing, and documentation procedures. This mirrors the quality control processes discussed in Chapter 5, but applies them specifically to the off-site context. Such requirements reinforce the idea that quality in MMC is not a single point in the process but must be embedded across design, manufacturing, transport, and installation stages. Procurement decisions that emphasise these criteria are therefore central to achieving the necessary end-to-end assurance.

It is also important to note that the integration of procurement and quality management requires a more collaborative approach between clients, designers, and manufacturers than in traditional projects. Because MMC involves the production of complete modules or large system components, any defects or misalignments can have significant consequences once delivered to site. Strong relationships and clear contractual mechanisms are therefore needed to ensure that quality standards defined at the procurement stage are carried through design development and into factory production. This is where procurement has a vital role, not only in setting minimum requirements but also in encouraging long-term partnerships with trusted suppliers who can deliver consistent performance across multiple projects.

Industry practice increasingly supports this integrated view, recognising that quality cannot be assured by price competition alone. Procurement strategies which consider certification, quality management processes, and supplier capacity alongside cost and programme are essential to the wider adoption of MMC. By embedding quality assurance into supplier selection and contractual frameworks, clients and contractors can mitigate risks, reassure stakeholders, and deliver outcomes that meet regulatory, safety, and performance expectations. In this sense, procurement is a powerful tool for ensuring not just that MMC is cost-effective, but that it delivers high-quality, safe, and durable buildings.

Case Study Analysis

Tide Construction. (2023). College Road. https://tideconstruction.co.uk/projects/college-road/

Case Study: College Road Development, Croydon – A Benchmark in Modular Construction

The College Road development in Croydon stands as a testament to the transformative potential of Modern Methods of Construction (MMC) in the UK. Completed in 2023, this project comprises two towers: a 50-storey structure with 817 co-living units and a 35-storey building offering 120 affordable homes. At 163 meters, it holds the title of Europe's tallest volumetric modular building.

Time Efficiency

One of the most striking achievements of the College Road project is its accelerated construction timeline. Utilizing a volumetric modular approach, 1,725 pre-fabricated units were installed in just 40 weeks, culminating in a total construction period of 28 months. This approach significantly reduced the typical construction duration, showcasing the time-saving advantages of MMC.

Cost Control

The strategic procurement of manufacturing slots and materials played a pivotal role in managing costs. Early engagement with suppliers and securing factory time ensured a steady production flow, mitigating potential delays and cost escalations. Additionally, the use of off-site manufacturing reduced on-site labour requirements, contributing to overall cost savings.

Quality Assurance

Ensuring high-quality standards was paramount throughout the project. The modular units were produced in a controlled factory environment, allowing for rigorous quality control processes. Adherence to certifications such as BOPAS and NHBC Accepts provided assurance of the durability and safety of the modules. Furthermore, factory audits and quality plans aligned with established QA practices were implemented to maintain consistent quality levels.

Design Freeze

A critical aspect of the project's success was the implementation of a design freeze early in the process. This decision ensured that the design was finalized before manufacturing commenced, preventing costly changes during the production phase. The design freeze facilitated a smooth transition from design to construction, aligning with the manufacturing schedule and reducing potential delays.

Warranties and Bonds

To safeguard the project's integrity and reassure stakeholders, comprehensive warranties and bonds were established. These financial instruments provided assurances regarding the performance and longevity of the modular units, addressing potential risks associated with off-site manufacturing. The inclusion of warranties and bonds reinforced the project's commitment to quality and reliability.

Contractual Arrangements

The unique nature of modular construction necessitated tailored contractual arrangements. Bespoke contract provisions were developed to address the specific requirements of off-site manufacturing, including early payments for factory production and off-site material insurance. These contractual mechanisms facilitated smooth coordination between stakeholders and ensured financial security throughout the project's lifecycle.

Future Developments

The success of the College Road development has set a precedent for future modular construction projects. The project's innovative approach and positive outcomes have garnered attention from industry professionals and policymakers alike. Lessons learned from this project are being applied to subsequent developments, influencing procurement strategies and industry practices.

Tender types

As with procurement, but also linked closely with the same process of selection, the same is true of tender methods and processes. This is the mechanical means by which the contractor provides the information to the client to enable contractor selection to happen effectively within the overall procurement route discussed above. There are several options, and some will only work with certain types of procurement, and others are more generic. The choice of which tender process to use is often a decision directed by the client advisors, typically the Quantity Surveyor or Project Manager, and the desire of the client themselves; However, as with procurement selection above, the emphasis on time, cost, and quality will also direct the choice. It is also worth noting that we keep using the term tender, rather than price – the reason for this is that in contemporary tendering price is not always the most important consideration and is rarely the only consideration when selecting a contractor.

Competitive Tendering

Typically, the tender system is split into competitive where more than one contractor is approached for a tender, and negotiated where just one is approached. Competitive tendering takes longer to process but can often provide a more desirable and fairer result, whereas negotiated tendering is often quicker, but as only one contractor is involved there is little possibility of gaining a wider perspective on the project.

Open Tendering

This is open to all, is often advertised in either specific locations or wider in general newspapers – the advantage of this type of tender is you can attract interest from sectors and companies you may not know exist or are interested. It can also help to broaden the consultants' knowledge of the state of the current contracting sector to help create selective tender lists. Disadvantages are that it takes much longer to analyse the tender returns as you do not always know the companies responding – it is also potentially more costly to send out documents as they can go much wider and farther than with more specific selective tendering.

Selective Tendering

Tender lists will often be used in a competitive tender situation – a tender list is a list that is created and maintained but one of the design team, often the quantity surveying practice or the architects practice, and this list of a few contractors are the organizations invited to tender. The starting point if one does not exist is often to use open tendering to discover the state of the contractors in that particular field – as MMC is still a growing market in the UK, and indeed in many parts of the world, this may be needed for any client or consultant wanting to use this production method. Once the potential contractors, who are capable of working in that sector (in the case of this book that is the MMC sector) are found, this can be narrowed down through a Pre-Qualification Questionnaire (PQQ) process. Specific important questions are asked of the contractors – such as experience, strength of management structure, financial strength, quality records, etc. – this allows a systematic narrowing of the number of potential contractors to a size of list that is manageable but will still give the required breadth of competition.

Once the tender list is developed, the principle is that these are the contractors that are approached to provide a tender for each of the projects – one disadvantage to this is that it becomes difficult to discover new contractors that may be better suited to a project, but an advantage is the speed with which tender analysis can happen. As the contractors are already scrutinised for capability, the specifics of the particular project can be focused on which makes it much quicker and less risky. This would be a typical tender route for projects involving MMC as there is a need to ensure the contractors are experienced in the processes used for production.

Single and Two-Stage Tenders

Single or two-stage tendering is a choice between focusing on cost, or on other aspects, in single-stage all information of the tender is presented and analysed at once. In two-stage tendering other quality aspects are focused on first, narrowing down the competition to one or two, and then the focus is placed on cost detail, sometimes turning into a negotiated second stage.

Negotiated Tendering

As mentioned above, negotiated tendering is where just one contractor is approached – this can place higher risks on costs of the project as when there is a monopoly there is often a choice, conscious or not, to take advantage and increase costs. However, sometimes there is no choice, and often this can be the case with MMC as there is not the wide choice of contractors, those who understand the system well enough to complete the project effectively. There are typically two reasons for opting for negotiated tendering: either the project is small and therefore not of great consequence in spending time looking for alternatives (consider domestic repairs, it is rare to search many different contractors when you need a door replacing at home); or when there is no competition available, there is only one contractor with the knowledge and capability to carry out the type of project. In the early days of MMC use, this is a possibility for some projects.

Partnering and Target Costing

The key principle of partnering is to create a process where the contractor and the client work together on the project from an early stage, with the purpose of creating an ideal outcome in all aspects of the project including financial. Using a target negotiated cost as the basis of this will enable decisions during the design and build process to be to the benefit of both parties, providing the best outcome. However, there were issues early on in the use of this type of system in that all of the pressures were passed down to the sub-contractors – with both client and contractors working together in collaboration, there was little protection for the sub-contractors. In a modern partnering arrangement, all key sub-contractors and suppliers are included in the agreement with the "pain and gain" philosophy running through the supply chain. If, by working together, the client and entire supply chain deliver the right quality under the target, all share that financial benefit. If, however, they do not work together and the target is missed, that loss is shared among all. One of the founding principles is that of developing long term relationships (if it works on this one, keep the team together and do the next project, etc.). This is a good principle to adopt with MMC, particularly if using volumetric systems as one of the key issues in MMC use is the lack of understanding between the factory production outputs and the on-site outputs – using partnering will allow for a quicker development of understanding and detail between all of the stakeholders, making future projects run smoother.

Chapter Summary

Procurement and tendering practices are central to determining the success of Modern Methods of Construction (MMC) in the UK. This chapter has explored how procurement is not merely a contractual process but a strategic mechanism for achieving the key outcomes of time, cost, and quality that underpin project success. By aligning procurement strategies with the distinct requirements of MMC, clients and contractors can maximise the benefits of off-site manufacturing while mitigating its unique risks.

A consistent theme throughout has been time predictability. Unlike traditional projects, MMC requires significant early planning, particularly through a design freeze that must be agreed before manufacturing can commence. Procurement is therefore the stage where clients must decide on the extent of MMC use and commit to early design certainty. When effectively managed, this enables factory slots to be secured, reduces delays, and ensures smoother sequencing of on-site and off-site activities. Inadequate consideration of these issues during procurement, however, can undermine delivery schedules, leading to time loss and disruption.

From a cost perspective, procurement plays a crucial role in balancing financial certainty with risk management. For instance, contracts that require manufacturers to begin production early, often necessitate upfront payments. Mechanisms such as vesting certificates, project bank accounts, and off-site material insurance are increasingly used to protect these investments while ensuring suppliers have the cash-flow to proceed. More broadly, procurement choices influence how risk is distributed between client and contractor: while lump sum tenders transfer risk to the contractor, they often result in higher prices; collaborative approaches, by contrast, can reduce risk premiums and support more accurate market pricing. Ultimately, the procurement strategy must strike a balance between minimising cost and ensuring commercial viability.

Quality assurance has also been shown to be deeply embedded in procurement. The selection of suppliers cannot be based on price alone: criteria such as BOPAS, NHBC Accepts, and ISO certifications provide confidence in manufacturing processes and product reliability. Procurement can further mandate factory audits, warranties, and long-term bonds, ensuring that off-site components perform to specification once installed on site. This is particularly important in a post-Grenfell context, where quality, compliance, and safety are inseparable. The discussion links directly to Chapter 5, where the detailed processes of quality assurance are explored more fully.

Contract structures also take on heightened significance in MMC. Traditional forms of contract were not designed with off-site manufacturing in mind, and as such procurement must adapt to include bespoke provisions for factory-based production, storage, and logistics. Clear terms around responsibility for defects, warranties, and long-term maintenance are vital in ensuring accountability across the lifecycle of MMC components. In this way, procurement safeguards not only immediate delivery outcomes but also the longer-term performance of the asset.

Finally, procurement is the mechanism through which future developments and innovations in MMC can be supported. Emerging approaches such as the platform-based model of construction rely on repeatability, standardisation, and aggregated demand across projects. By procuring standardised components at scale, clients and government can create the stable pipeline that manufacturers need to invest in capacity, which in turn improves time, cost, and quality outcomes across the sector. Initiatives such as the Construction Playbook and the Infrastructure Projects Authority's promotion of platform approaches demonstrate the UK government's recognition of procurement as an enabler of MMC at scale.

In summary, procurement is not an administrative hurdle but a strategic tool. It directly shapes the outcomes of time predictability, cost control, and quality assurance, while ensuring that critical issues such as design freeze, risk allocation, warranties, and future scalability are addressed. When procurement is aligned to the realities of MMC, it enables the industry to realise the benefits of faster delivery, controlled costs, and improved quality – offering a clear route to the more sustainable and efficient construction practices outlined in the chapters that follow.

References

Cabinet Office (2020). *The Construction Playbook: Government Guidance on Sourcing and Contracting Public Works Projects and Programmes*. London: Cabinet Office. www.gov.uk/government/publications/the-construction-playbook (accessed 25 August 2025).

IPA (2021). *IPA Blog*. https://ipa.blog.gov.uk/2021/11/19/all-aboard-platforms-for-buildings/ (accessed 8 August 2025).

Chapter 4

Supply Chain Management

Introduction

Supply chain management lies at the heart of construction, serving as the critical link between planning, procurement, production, transportation, and on-site assembly. When well-managed, it allows projects to progress with fewer surprises and more consistent outcomes in terms of time, cost, and quality. However, when the supply chain falters, the consequences are often delays, increased expenses, and declining standards.

The traditional model of construction supply chains has served the industry for decades. Typically, multiple parties come together, often for a single project, and coordinate their work through meetings, instructions, and a foundation of professional trust. Long-term relationships and clear lines of communication can help maintain order, but the sheer number of moving parts often makes the process feel fragmented and difficult to manage.

The increasing adoption of modern methods of construction (MMC) is transforming this picture. More work is now carried out off-site in controlled factory environments. Decisions must be made earlier in the process, and what arrives on-site is no longer just raw material, but prefabricated products and systems. This shift opens up new opportunities to enhance quality, predictability, and efficiency, but only if the supply chain is designed deliberately and managed as a continuous flow from design to handover. In other words, the chain must be actively shaped; it cannot be left to assemble itself through habit or default.

To support this, Chapter 2 outlined the MMC categories that underpin the approach discussed here, while Chapter 3 explores the procurement choices that determine who does what, and when. This chapter builds on both, showing how supply chain management transforms strategic intent into effective delivery.

Our focus here is practical. We examine which aspects of the traditional approach still serve us well, and which ones fall short when applied to MMC. We explore how early supplier engagement, the use of platform-based components, and timely design decisions help to establish a smoother, more reliable flow of work. We also highlight how digital tools enable teams to identify potential clashes, schedule lifts, and coordinate logistics before site work begins.

Crucially, we treat transport and storage not as gaps in the process, but as key stages in safeguarding quality. We draw on the concept of the golden thread (introduced in Chapter 5) to show how evidence and accountability can be maintained seamlessly from factory floor to final crane lift. We also link forward to Chapter 6, which demonstrates how time can be saved through earlier planning and the use of clear programme gateways. Finally, we preview Chapter 7, where a predictable flow of work contributes to stronger cost control and commercial certainty.

DOI: 10.1201/9781003570431-4

In essence, supply chain management is the mechanism through which good plans become reality. It is far more than a schedule of orders and delivery dates, it is the underlying structure that connects design freeze, factory slots, transport windows, and site readiness into a coherent whole. When executed well, it enables MMC to fulfil its promise; when neglected, the potential benefits quickly unravel. This chapter outlines the key steps needed to build a supply chain that is reliable, transparent, and fit for repeatable, high-quality delivery.

Historical Perspective

Where the Traditional Model Came From

The traditional construction supply chain developed in response to a need for reliability, trust, and local familiarity. Projects were typically commissioned one at a time, with clients engaging designers to create bespoke solutions tailored to the specific requirements of each job. Main contractors would then assemble delivery teams by drawing on established networks of subcontractors and preferred suppliers, relationships built and strengthened over many years. This approach fostered a relatively stable way of working.

Information moved in a linear sequence: design came first, followed by procurement, then construction. The system rewarded practical, hands-on skill and the ability to adapt to problems as they emerged on site. It is this combination of craft and adaptability that helped the traditional model serve the industry so well for so long.

The diagram of the traditional flow illustrates this origin clearly. Distinct lanes represent who is responsible for what, and when, while blue arrows show the planned progression from client brief to project handover. This visual reinforces the core logic of a model built around sequential handovers and well-understood professional boundaries.

At its core, the traditional supply chain is defined by a project-by-project, trade-by-trade approach. Each job is treated as unique. Designers produce custom drawings and specifications for each site. Main contractors let work packages to specialist subcontractors, and suppliers respond to purchase orders only when the project is ready to receive materials. Deliveries are typically timed to arrive just before installation, and quality is assessed on site by experienced teams able to physically inspect the work.

Construction progresses through a recognisable sequence: groundworks, structure, envelope, services, finishes, testing, and finally handover. Each stage typically waits for the previous one to be completed before beginning. The sequence is maintained through meetings, site instructions, and regular inspections. When unexpected changes arise, the team resolves them on site, relying on coordination, improvisation, and supervision to keep the project moving.

This system has clear strengths. It is highly flexible at the point of delivery, with skilled trades able to respond to real-time conditions, whether those are changes in the weather, unforeseen site constraints, or evolving client preferences. Minor adjustments, such as moving a wall or re-routing a pipe, are often managed with minimal delay. Longstanding supplier relationships add further resilience: a trusted ready mix plant or steel fabricator can help smooth over supply issues and maintain momentum.

There is also a visibility benefit. Because most construction activities occur in the open, site managers can observe quality firsthand, identifying and correcting issues before they are concealed by subsequent trades. Chapter 5 returns to this point when comparing the role of in-field inspection with factory-based quality assurance.

Structural Limits That Show Up at Scale

However, the same characteristics that give the traditional model its flexibility also reveal its limitations, especially on larger or more complex projects. The chain is fragmented across multiple firms, with each discipline operating largely within its own silo. Information flows between them via drawings, emails, and formal meetings, creating points of friction and risk. When details are unclear or incomplete, the default response is to resolve them on site, rather than upstream at the source. This leads to delays, rework, and inefficient resequencing of tasks.

Quality becomes harder to control as external variables multiply. Work is exposed to weather, and project schedules are vulnerable to rain, frost, or wind, each of which can disrupt progress and compromise finish quality. Labour is often mobile, and skilled trades may not be available at key moments. Materials are ordered per project and frequently stored on site, increasing congestion, the risk of damage, and logistical complexity.

These structural challenges are visualised in your diagram by the red feedback loops and the orange risk points clustered around storage areas, trade interfaces, and crane re-bookings.

The result is a system that absorbs delay rather than preventing it. Time buffers, or "float", are often hidden within the sequence: a late drawing delays a purchase order, a late order delays a delivery, and a late delivery holds up a trade. Experienced managers can keep the job on track, but often at the cost of predictability. Chapter 6 explores this further by comparing traditional, sequential programmes with more modern, parallel approaches.

Why Modern Pressures Strain the Model

Today's construction projects are subject to a new and more demanding set of pressures. Urban sites often have limited access and tighter working windows. Clients now expect faster completions with minimal disruption to surrounding communities. At the same time, building regulations require improved fire, acoustic, and thermal performance, supported by robust, verifiable evidence. Carbon reduction targets demand fewer deliveries, reduced waste, and more standardised, repeatable details. Digital coordination has also become the norm, increasing expectations for precision and integration.

The traditional supply chain is capable of meeting these challenges, but it does so with increasing effort, cost, and complexity. Designs remain highly bespoke, even where repetition might offer benefits. Lessons learned are not always transferred from one project to the next because delivery teams frequently change. Procurement remains reactive, fragmented across multiple packages. And site-based improvisation is still the primary tool for absorbing late-stage change.

Lessons We Carry into MMC

The aim is not to discard the traditional model, but to build on its strengths while addressing its limitations. Many aspects of traditional practice continue to deliver value: long-term supplier relationships remain important, clear site-level accountability is essential, and statutory checks by building control are still a core part of project assurance.

What changes with MMC is the method of delivery. More decisions are made earlier in the process. Standardised components and platform-based approaches reduce variation and support repeatability. Design is aligned to known grid sizes and riser zones. Product passports are used to ensure that every component carries traceable evidence from factory to crane. Transport and storage are no longer treated as logistical afterthoughts, but as key quality stages within the overall delivery process. The golden thread of records, introduced in Chapter 5, is maintained in a shared digital space, making the full history of each component visible from design through to installation.

Chapter 6 presents the MMC version of the traditional process diagram. It includes a dedicated factory lane running in parallel with site work, connected by well-defined gateways that coordinate both streams. Chapter 5 sets out the specific check points, certificates, and approvals that demonstrate each step has been completed as intended.

Together, these chapters illustrate how the lessons of traditional construction, especially its emphasis on relationships, supervision, and hands-on skill, can be integrated into a more predictable and scalable flow of work.

We have now established the baseline. The traditional supply chain thrives on trust, craft, and responsiveness. But it becomes strained under the weight of modern expectations for speed, consistency, and evidence. The next sections explore how MMC reshapes this picture, transforming materials into products, using platform thinking to reduce variation, engaging suppliers early to support design freeze, and planning the entire chain as a single, integrated system. In short, we move from a well-worn sequence to a better-designed flow.

Practices That Have Remained Unchanged – And Why

Despite significant changes in construction methods and technologies, several core practices within the construction supply chain have remained largely unchanged, and for good reason. Long-term relationships between contractors and suppliers continue to provide a foundation of reliability. These relationships are built on shared history, mutual understanding, and a track record of delivery, all of which contribute to more consistent outcomes on complex projects. Trust, once established, reduces the need for over-specification and close policing, allowing teams to focus on delivery rather than dispute.

On-site quality control also remains an essential feature of construction delivery. Supervisors and trade leads are able to inspect work in real-world conditions and request adjustments in real time. This responsiveness is one of the enduring strengths of site-based production, it allows teams to react quickly to external factors such as weather, access constraints, or unforeseen clashes between systems. In this context, flexibility and immediacy are not just operational conveniences; they are core enablers of progress.

The familiar sequence – design, then procurement, then construction – still underpins the structure of most projects. Figure 4.1 illustrates why this linear model continues to be widely used. It offers an orderly way to coordinate many different specialist skills, ensuring that each stage reaches an acceptable standard before the next one begins. Although digital tools have accelerated processes such as drawing exchange and issue tracking, the underlying sequence has endured because it offers a simple, shared framework for managing complex work across diverse teams.

The Enduring Role of Skilled Trades

Skilled trades remain central to construction delivery. Carpenters, electricians, plumbers, and other site specialists carry hard-won craft knowledge and contribute problem-solving ability at the point of work. Their expertise introduces flexibility into the system, enabling teams to adapt, interpret design intent, and resolve minor issues without escalating delay or cost.

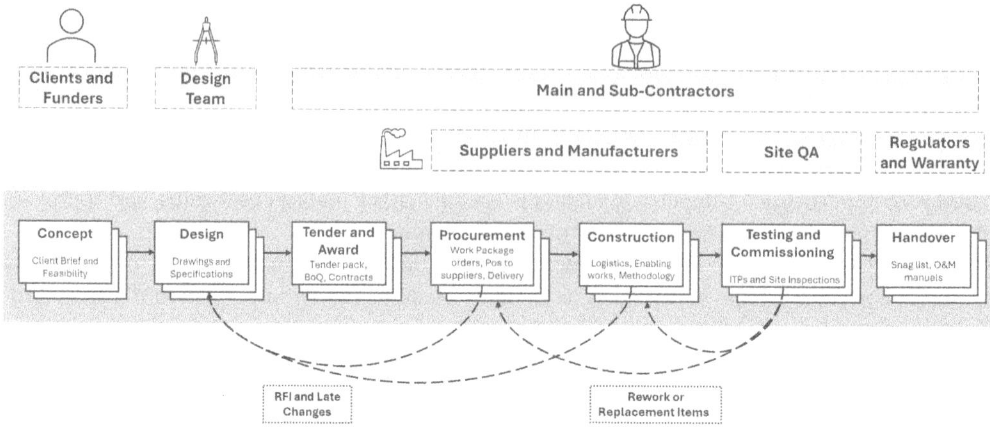

Figure 4.1 The Core Logic of the Traditional Chain

While automation, prefabrication, and modern methods of construction (MMC) are increasingly prevalent, they do not replace the judgement and adaptability of experienced tradespeople. In many cases, their presence ensures that quality is maintained through attention to detail and practical intuition. These long-standing practices continue to anchor the industry, providing stability and predictability during a time of broader change.

Persistent Weaknesses in Procurement Strategy

Where the traditional chain has often struggled is not in the quality of its people but in the absence of a strategic approach to procurement. Many projects continue to operate without a joined-up view of purchasing. Orders are placed on a project-by-project basis, and relationships are frequently rebuilt from scratch for each new job. This fragmented approach misses opportunities to agree on shared standards, carry forward learning, or negotiate better commercial terms across a pipeline of work.

The consequences can be significant. Costs may rise due to missed economies of scale. Material quality can vary between packages or projects. Lead times may be longer or less predictable. Without a strategic framework, procurement becomes reactive, increasing the risk of inconsistency and reducing the scope for innovation.

Gaps in Collaboration

Collaboration in procurement has also been uneven. By collaborative procurement, we mean arrangements in which key parties, clients, designers, contractors, and suppliers, agree on project outcomes together and share information early. Examples include two-stage design and build contracts with early supplier input, frameworks that incorporate standardised details, or alliances built around common goals for time, cost, and quality.

When this level of collaboration is missing, each firm tends to focus on its own package in isolation. This siloed thinking leads to poor coordination: one team may over-order materials to hedge against risk; another may plan work based on crew availability rather than site readiness. These disconnected decisions lead to duplication, inefficiency, and blurred responsibility.

A common example is a riser designed to a certain specification, only to be rerouted during MEP coordination and then modified again on site to suit a panel system no one had accounted for. While such issues can often be resolved, they consume time, cost money, and reduce confidence in the wider system.

Weak strategic planning and limited collaboration also create knock-on effects in logistics and sustainability. Sourcing from multiple suppliers without regard to location or delivery timing can increase transport distances, handling requirements, and carbon emissions. Uncoordinated delivery schedules create congestion and heighten the risk of damage during site storage. In contrast, better planning could consolidate loads, reduce trips, and make use of local manufacturers where appropriate.

Building a Bridge to MMC

This is a useful point to reintroduce MMC. The methods discussed in later chapters, such as platform-based parts and repeatable assemblies, are designed to tackle precisely the issues outlined above. Standard interfaces reduce the need for on-site rework. Early design freeze creates a single version of the truth, reducing variation and avoiding late change. Fewer, larger, and better-planned deliveries simplify logistics and improve handling.

These improvements are not intended to replace the traditional model outright, but to augment it. The aim is to retain what works, long-standing relationships, visible site responsibility, and the skills of the trades, while adding new practices that support strategic, collaborative planning. In doing so, the supply chain can operate not as a collection of disconnected parts, but as an integrated system with a shared goal.

The Impact of Urbanisation

Urbanisation, by which we mean the increasing concentration of construction activity in dense city environments, has reshaped the practical realities of project delivery. In these urban settings, space is limited, neighbouring buildings are close, and regulations governing traffic, noise, and safety are far more stringent. Whether it's a hospital campus, a city-centre station, or a multi-block housing development delivered over several years, urban projects bring complex programmes and high public visibility.

These conditions place new pressures on the supply chain. Projects must now be more predictable, quieter, safer, and faster to deliver on site. The traditional practice of ordering loose materials and assembling everything piece by piece on the street is no longer viable in many urban contexts.

In response, the industry has begun to shift away from bespoke, project-specific parts toward more standardised products, designed once, manufactured in batches, and deployed across multiple projects. This transition is supported by longer-term commercial relationships such as frameworks and alliances, which provide a stable pipeline of work for manufacturers and more consistent outcomes for clients. As discussed in Chapter 3, these procurement arrangements also reshape how risk and timing are managed. Chapter 6 builds on this by showing how earlier planning converts that consistency into greater programme certainty.

Standardisation for Urban Efficiency

In city environments, the use of standard and repeatable elements simplifies delivery. Rather than inventing new solutions for every detail, teams increasingly draw from a library of tested components with known dimensions, fixings, and tolerances. These elements can be small or large in scale, ranging from a prefabricated bathroom pod to a façade cassette or a simple timber roof truss.

The timber truss provides a helpful example, particularly for first-year students. Factory-made roof trusses are now common across UK housing. They arrive on site with connection plates fixed, dimensions checked, and lifting points clearly marked. As a result, the site team can install them in a predictable rhythm, using fewer people and generating less noise than would be required for a traditional "stick-built" roof. The benefit is not just speed, it is steadiness, predictability, and reduced site disruption.

This is the essence of the shift from a project-centred, one-off approach to a more industrialised, repeatable process. Crucially, this does not mean losing design freedom. Instead, it means establishing a platform of standardised parts and interfaces, and varying the layout, scale, or finish within that platform. Chapter 2 introduced this concept of platform thinking in more depth. The key advantage is that suppliers can manufacture in batches, quality checks become repeatable, and installation methods can be learned once and applied many times, reducing variability and increasing efficiency.

Outcome-Focused Manufacturing

This movement toward standardisation is also matched by a broader shift in emphasis from inputs to outcomes. Where previously a project might begin with a material list and a bespoke detail, the starting point is now increasingly the performance required. Design and procurement decisions are based on tested outcomes such as energy efficiency, fire safety, acoustic performance, durability, and ease of maintenance.

For example, a wall system might be selected not simply for its material composition, but because it delivers a certified U-value for thermal performance, a known fire resistance rating, and verified acoustic separation. Once selected, that system is then manufactured repeatedly to the same specification, ensuring the on-site result matches the tested performance.

This outcome-focused logic runs throughout the supply chain. Manufacturers invest in jigs and repeatable processes to maintain tolerances. They create product passports, digital or physical documents that provide evidence of what each unit contains and how it was tested. Designers align their drawings to these known interfaces, reducing unnecessary variation. Clients and funders gain confidence from the traceability and repeatability of delivery. Chapter 5 explains how Quality Inspection and Test Plans (QITPs) and Certificates (QITCs) help maintain this golden thread of evidence across each stage. Later in this chapter, we also connect these choices to carbon metrics and to environmental rating systems such as BREEAM and LEED, where product declarations and lifecycle assessments support transparent reporting.

Logistics as a Design Challenge

Urban construction logistics present a unique challenge. Sites often have minimal laydown space, narrow access routes, and tightly controlled delivery windows. In this context, standardised parts enable precise lift planning and reduce the time spent at the crane hook. Fewer deliveries mean fewer disruptions to neighbours, less noise, and a lower risk of materials being damaged due to lack of space or improper storage.

In some cases, teams can take advantage of a near-site making space or a flying factory, a temporary production or assembly facility located close to the project. These allow subcomponents to be assembled locally, reducing transport distances and providing a safe area to test interfaces before lifting them into position. We return to the topic of flying factories later in this chapter when we explore supplier models in more detail.

While repeatable design and streamlined logistics improve delivery, they do not remove the need for buffer planning. Even the best-laid programmes must account for unexpected delays due to roadworks, bad weather, or the need to rebook a crane. Chapter 6 addresses how buffers are built into modern programmes and how they differ from traditional float.

Standardisation also requires professional judgement. Some tight urban sites may restrict the size of modules that can be delivered. Other projects may require bespoke solutions to meet planning conditions or heritage requirements. The challenge is not to standardise everything, but to find the right balance between repeatability and context-sensitive design. In other words, we standardise where it helps, and customise where it matters, retaining the value of traditional craft and contextual response, while introducing predictability through platforms and repeatable assemblies.

A Shift in Mindset

Urbanisation has fundamentally changed the conditions under which construction takes place. It has nudged the industry away from site-based improvisation and toward product-based solutions. It has made early decisions more important, turned logistics into a design priority, and elevated the role of long-term supplier relationships.

The result is a more systematised, outcomes-driven approach to construction, one that works particularly well in dense urban contexts where space, time, and public impact are critical considerations.

In the sections that follow, we explore the digital tools, early trials, and hybrid delivery models that support this shift. We look at how small-scale or phased adoption of MMC can offer practical entry points for projects. We also examine how transport and storage are no longer treated as passive steps, but as integral stages of quality control. Throughout, we will continue linking to Chapter 5 on evidence and Chapter 6 on programme logic, reinforcing the principle that the supply chain must operate as a joined-up system, not a series of isolated tasks.

Digitalisation and Early-Stage Trials

Digital tools are transforming how projects begin, enabling early-stage trials that shift critical decision-making from the site to the design phase. Decisions that once occurred during construction, often under time pressure and at higher cost, can now be explored and tested virtually. This change reduces rework, supports value engineering, and allows teams to pre-select products and systems that align with performance targets and client goals.

Building Information Modelling (BIM) sits at the heart of this shift. It brings multiple disciplines into a single, shared model, enabling designers, engineers, and contractors to coordinate in three dimensions. Clashes between structural, architectural and MEP systems can be detected and resolved early. With programme data added, teams can simulate the construction sequence in four dimensions, testing logistics, identifying bottlenecks, and confirming site readiness ahead of time.

The model also becomes a platform for evaluating cost, carbon, and performance. Options can be assessed not just by price, but by value, considering lifecycle cost, embodied carbon, and long-term efficiency. This supports clearer, earlier design freezes and more confident engagement with suppliers.

Beyond visualisation and coordination, digital models can be used to simulate building performance. Thermal modelling, daylight analysis, acoustic testing, fire strategy evaluation, and structural behaviour can all be explored before a single order is placed. For example, wall types and window specifications can be adjusted to achieve a target U-value, enhance airtightness, and improve comfort in use. Services can be virtually installed to test for access, maintenance pathways, and eventual replacement, all before anything is physically built.

Better Information, Smoother Delivery

These digital trials improve information quality across the supply chain. Product data can be linked directly to the model, and logistics can be rehearsed, checking whether elements fit through site gates, whether cranes can reach, and whether lifting points are in the right place. A common data environment (CDE) stores design decisions, specifications, approvals, and records in one accessible space, ensuring continuity from design through factory and on to site.

This traceability supports better quality control, smoother handovers, and stronger assurance at every stage.

However, early digital trials only deliver value when they are built on reliable foundations. Teams need high-quality survey data, agreed modelling standards, accurate manufacturer inputs, and simple, agreed acceptance criteria for each decision. With these in place, digital design becomes a safe environment to test and learn, de-risking delivery and supporting a more predictable, efficient build process.

Supporting Standardised Elements

Standardised elements are manufactured products and sub-assemblies that share fixed interfaces, defined tolerances, and repeatable methods of production. These might include factory-built bathroom pods, façade panels, wall cassettes, pre-assembled risers, or services modules. The goal is not uniformity for its own sake, but stability: using a common kit of parts across multiple projects, with controlled variation where it adds value.

This shift turns attention from branded products to performance outcomes. Elements are selected based on how well they deliver against measurable criteria, such as durability, airtightness, thermal and acoustic performance, fire resistance, embodied carbon, and maintenance requirements. This focus on outcomes improves predictability, reduces waste and rework, and strengthens control over time and cost.

Digital Tools Make Repeatability Possible

Digitalisation plays a key role in making standardisation work. A shared BIM environment holds all the relevant data, design geometry, product specifications, and installation rules, in one coordinated model. Buildability can be tested virtually, including lifting strategies, access for maintenance, and sequencing. Cost and carbon comparisons can be run quickly using product data templates and environmental product declarations (EPDs), giving teams a common language for assessment and evidence.

Early digital trials support earlier design freeze, protect interfaces, and give factories the stability to plan repeatable manufacture. When models are well developed, information flows cleanly from design to factory to site, and handover becomes a continuation of that process rather than a separate phase.

Aligning Procurement and Delivery

Procurement practices must adapt to support standardised delivery. Rather than naming a preferred brand, specifications focus on performance outcomes and the data required to prove them. This opens the door to more flexible sourcing while maintaining standards.

Early supplier involvement is key. It allows manufacturers to plan tooling, build jigs, and allocate production slots with confidence. It also helps logistics teams prepare for packaging, weather protection, delivery timing, and booked lifts. With input from the factory, site teams can rehearse how elements will be installed and where temporary works may be needed.

Quality control benefits from clearer structure. Agreed acceptance criteria, planned hold points, and a single non-conformance process remove ambiguity. A firm change control process protects the product configuration. Maintaining pre-approved alternatives for critical components adds resilience in the face of supply constraints.

Training, mock-ups and sample units help align factory and site teams, ensuring that what is designed and manufactured can be installed efficiently and to the right standard.

Together, these practices allow the supply chain to deliver repeatable, high-quality components that work first time, and that support the larger shift toward platform thinking, evidence-based assurance, and predictable delivery.

External Factors and Their Impact

External factors are forces beyond the control of individual projects but which have a significant influence on how construction supply chains operate. These include shifts in policy, market conditions, and global events. Each can impact the selection of products, the timing of procurement, delivery risk, and overall project cost. In this section, we examine key pressures: globalisation, macroeconomic shocks, regulatory change, pandemics, energy price volatility, geopolitical events, climate disruption, labour shortages, logistics constraints, digital risks, and insurance conditions. Understanding these external dynamics allows teams to respond with appropriate controls, maintain resilience, and deliver with greater predictability.

Globalisation Increases Trade Exposure

Global sourcing has expanded the range of available products and suppliers, giving projects access to innovation and competitive pricing. However, it also introduces new risks. Prices shift with currency movements. Transport routes are longer, which means delays accumulate more easily. Customs regimes differ by country, and this brings complexity in the form of additional paperwork, inspections, and fees. Ethical sourcing also requires clearer proof of origin, backed by audits and certification.

In response, the industry increasingly relies on fixed interfaces and pre-approved alternates to avoid redesign if a supplier changes. Procurement strategies are adapting by pre-qualifying more than one source, defining fair mechanisms for adjusting prices, and applying tools such as vesting and project bank accounts to improve payment security. These measures help build resilience into the system without sacrificing flexibility.

Macroeconomic Shocks Tighten Finance

Wider economic shocks, such as the 2008 financial crisis or more recent interest rate rises, have had a constraining effect on project finance. Appetite for risk reduces, and cash flow becomes more fragile. In these contexts, standard site-based delivery struggles to absorb delay or uncertainty without cost escalation.

A more robust response includes gated approvals, staged payments tied to verifiable progress, and clearer commercial models that anticipate earlier spend on off-site manufacture and reduced site prelims. Smaller programme buffers, aligned with readiness checks at project gateways, help to protect dates even when financial pacing shifts. The key is to ensure that decision-making remains active and clearly owned, even when external pressure rises.

Geopolitics and Regulatory Change

Geopolitical events can reshape delivery routes and procurement options almost overnight. Sanctions, conflict, or the closure of major transport channels can alter which ports are accessible, how long shipments take, and what it costs to insure them. Containers may be delayed, rerouted or repriced. Certain product certifications may no longer be accepted, and suppliers may pull out altogether. Brexit illustrates how these dynamics play out in practice, introducing new recognition paths for construction products, increased customs checks, and labour shortages across the trades. The path from factory to site has become slower, more fragmented, and less certain.

To counter these risks, the industry must bring compliance tasks forward. This includes confirming accepted standards early, comparing international specifications with domestic regulation, and scheduling any retesting before orders are placed. Planning should extend beyond purchase dates to include route selection and logistics permits. Near-site consolidation points or regional storage can reduce final-stage exposure. Delivery itself becomes a quality stage, with packaging, weather protection, and scheduled lifting factored in. Arrival checks must verify not only condition but documentation. As discussed in Chapter 5, keeping all test reports, certificates and audit records in a single, trusted system preserves the golden thread.

One way to reduce exposure is by developing regional partnerships and engaging suppliers closer to site. This shortens transport, improves resilience, and often supports environmental goals. Commercial mechanisms such as vesting and agreed adjustment clauses can protect cash flow. As part of that response, the industry must also look to its future workforce. Geopolitical disruption reduces mobility. We must therefore invest in local training and apprenticeships to build installation and commissioning capability in areas where products are delivered. In short, geopolitical change alters the shape of the delivery route. We must in turn alter the sequence of checks, build in greater transparency, and grow capacity where it can be controlled.

Pandemics Disrupt Factories and Sites

Pandemics, such as COVID-19, halt the steady rhythm of construction. Factory shutdowns, transport limitations and lockdowns on site all led to a breakdown in established sequences. Even basic tasks became more difficult when people, vehicles, and materials could not move freely.

Projects that responded most effectively were those with a visible, flexible plan. A shared calendar between factory and site created a joined-up view of tasks. Sequencing tools allowed teams to model what could shift and what could not, introducing buffers where they made the most difference. These methods are discussed further in Chapter 6. At the same time, assurance had to adapt. Remote inspection, photographic evidence and checklists helped maintain confidence when travel was restricted, as outlined in Chapter 5. Where contracts recognised off-site milestones and set clear rules for access, decisions could be made more fairly. In uncertain conditions, projects benefit from simple rules, high-quality information, and the ability to adjust without losing sight of overall goals.

Energy Price Volatility

Sudden changes in energy prices have a direct effect on construction inputs. Core materials such as steel, cement, and glass become more expensive to produce and transport. Factory operations may introduce surcharges or scale back output. Site costs rise too, due to increases in fuel, heating and temporary power. Price quotes expire more quickly, delaying orders and shifting budgets.

In this environment, energy must be treated as a design and procurement input, not just a background cost. Design options can be compared based on their energy intensity, and procurement plans can include forward orders for items with strong justifications. Contracts should allow for price adjustments linked to clear, published indices, with transparent evidence at payment stage. Commercial leads must monitor trends in energy pricing and supplier alerts, making this movement visible in the five-dimensional model introduced in Chapter 7. In this way, volatility in energy prices becomes a managed factor, not a source of surprise.

Climate Events Test Resilience and Compliance

Extreme weather events, floods, storms, and heatwaves, are becoming more frequent and more disruptive. They close roads, ports and depots. They reduce safe working hours, interfere with curing, and affect adhesives, sealants and fixings. Products are more likely to be damaged in transit or during storage. As a result, programme reliability falls, and quality risk increases.

We should treat climate not as background noise, but as a design, logistics, and site control factor. Product selection should consider performance under known environmental conditions, using environmental product declarations and life cycle evidence. Sequencing must place sensitive tasks, hot works, wet trades, and heavy lifts, within safe, weather-aware windows. Storage plans must protect condition through appropriate cover, layout, and tracking. At arrival, checks must include temperature and moisture records, not just visual inspection. These principles are embedded in the quality system outlined in Chapter 5. Standardised interfaces and repeatable construction methods, as discussed in Chapter 2, support flexible rescheduling when weather windows shift. In other words, predictability comes from designing for variance.

Labour and Skills Gaps Reduce Capacity

Shortages in skilled labour reduce capacity across supervision, trades, and specialist tasks. When teams are stretched, installation slows, error rates rise, and inspection failures become more common. The resulting delays increase costs and place greater stress on the rest of the supply chain. Even when new personnel are brought in, productivity takes time to build. Where competence is not clear, approvals stall and payments drift.

To manage this, teams should reduce complexity at the point of install. Standardised assemblies, with generous access and simplified fixing, make tasks easier to teach. Training should be practical and visible, using mock-ups, sample units, and clear instructions. Readiness checks should test competence before new tasks are undertaken. Training time must be built into both the factory and site calendars so that planning reflects reality. Where appropriate, some work can move into factory settings where structured learning and quality oversight are easier to manage. These ideas are carried forward in Chapter 6.

Cyber Risks

As the digital tools that support design, procurement, and delivery become more central, the risks of cyber failure grow. System outages, corruption or ransomware attacks can block access to key files: models, schedules, product data, transport plans. Wrong versions of drawings or installation instructions can result in mis-builds and costly rework. If records are lost or incomplete, gateways cannot be passed and compliance cannot be proven. The result is delay, mistrust, and cost.

The response is to treat information management as a critical control. BIM models and common data environments (CDEs) must be governed with discipline. Standards for naming, versions, and approvals must be followed. Backups must be scheduled and restore processes tested. Critical documents must be held offline for essential site tasks. Gate checklists should confirm that the correct version of each document is in use. This protects the golden thread described in Chapter 5. Shared calendars across disciplines help to spot clashes early and maintain a coordinated programme, as discussed in Chapter 6.

Insurance Conditions Influence Behaviour

In the aftermath of loss events, insurance conditions become more stringent. Premiums increase, exclusions expand, and underwriters demand stricter evidence of control. If the project team cannot demonstrate effective risk management, cover may be reduced or withdrawn. Claims may be delayed, placing additional pressure on cash flow.

To avoid these outcomes, we must design for insurability. Roles must be clearly defined. Gates must be aligned to evidence. Quality Inspection and Test Plans (QITPs) and Certificates (QITCs) set expectations for performance at each stage. Test reports, photos, certificates and sign-offs must be stored centrally and consistently. As discussed in Chapter 5, the history of the project must be complete at handover. Insurers and funders should be engaged early, with a concise data room that shows how risk is controlled, how the supply chain is monitored, and how changes are documented. The commercial value of these choices is further explored in Chapter 7.

External factors rarely act in isolation. More often, they interact, one pressure triggers another, and the project must absorb the effects in sequence. The industry's response is to land the impact clearly, then lock controls into place through well-defined roles, decision gates, workflows, and traceable records. This is what allows us to build in confidence, even when the wider context is uncertain.

Benefits and Challenges of Small-Scale MMC: Cost, Quality, and Risk

Small-scale MMC refers to the use of selected, repeatable components that are produced in controlled off-site environments, while the main structure of the building follows a more traditional route. It offers a measured step towards modern methods of construction without requiring the full commitment of a factory-based programme. The focus is on elements that can be prefabricated, transported safely, and installed efficiently without altering the core structural design, examples include partitions, service risers, bathroom pods, or lightweight façade panels. This approach complements the method overview introduced in Chapter 2.

The primary consideration in small-scale MMC is cost. Initial investment is significantly lower than that required for full-scale manufacturing, making it a more accessible route for teams looking to test new methods without engaging in large capital programmes. However, this comes with a trade-off. Without the efficiencies of scale, unit costs may appear high and savings can be harder to quantify in the early stages. These early trials should be seen not simply as a delivery tactic, but as a form of strategic research and development. The value lies in reduced rework, fewer defects, improved programme certainty, lower material waste, and the development of a learning curve that can benefit future projects. For this reason, any pilot should begin with a clearly defined scope, agreed exit conditions, and a trigger for scale-up, such as a target installation time or a defect threshold. Budgeting tools that support both capital planning and life cycle return are discussed further in Chapter 7.

Quality is often where small-scale MMC delivers the earliest returns. A stable and repeatable factory process allows for more consistent outputs. Inspections are easier to carry out, and installation methods can be rehearsed in advance of the first delivery. To achieve this, the approach must be formalised with clear drawings, fixed jigs, and well-defined access conditions. Running a sample unit through a full first article inspection builds confidence in the method. Quality Inspection and Test Plans (QITPs) and Quality Inspection and Test Certificates (QITCs) define the key checkpoints, from material intake through production, packaging, transport, and site acceptance. Test reports and product data complete the record. This structure makes integration risk more visible. By isolating and proving individual components, such as a riser, partition system or bathroom pod, in a controlled sequence, the project team can identify and resolve clashes or access issues before they affect the wider programme.

However, risk emerges when moving from pilot to full deployment. A solution that performs well at small scale, with short transport distances and a focused crew, may not hold up under increased volume or tighter deadlines. Suppliers often prefer predictable demand and steady pipelines; sporadic orders can lead to higher prices or limited availability. Issues such as minimum order quantities, upfront deposits, and change penalties can quickly erode the perceived value of MMC unless planned for in advance.

To mitigate these risks, both design and procurement need to support scalability. Design should maintain fixed and clear interfaces so that multiple suppliers can work to the same connection rules. This keeps sourcing options open as demand grows (see Chapter 2). On the procurement side, a simple framework agreement with scheduled call-offs, transparent lead times, and evidence-based payment terms provides the necessary flexibility and control (see Chapter 3). In regions where the local MMC market is still developing, imported systems may be used to fill the gap, albeit often at a higher cost. In such cases, a hybrid delivery model can be employed: conventional methods are retained for low-volume or non-standard elements, while off-site production is focused on the few parts that provide the highest return on effort.

Operational planning determines whether a pilot is worth scaling. The workflow must include defined gates, design freeze, factory acceptance testing, logistics rehearsal, and site acceptance checks. Clear roles are essential: the design lead is responsible for interfaces; the factory lead manages jigs and process control, the logistics lead oversees protection and timed delivery; the quality lead manages evidence and non-conformity processes. The planner coordinates a joint calendar across factory and site so that installation windows are real, not theoretical, an approach further explored in Chapter 6. If the pilot fails to meet its targets, the team can step back or revise the method with minimal sunk cost. If the targets are met, then the model can be scaled by increasing volume, onboarding additional suppliers, or expanding the scope of off-site components.

Ultimately, small-scale MMC is a learning tool with bounded risk. The aim is to invest just enough to prove a reliable, repeatable method, to maintain flexibility through well-protected interfaces, and to embed quality into the process through the use of QITPs and QITCs. This approach not only supports supplier and logistics planning, treating delivery itself as a quality stage, but also lays the foundation for the tighter quality and time controls discussed in Chapters 5 and 6. It is a bridge between traditional site work and industrialised construction, offering the opportunity to experiment safely and scale wisely.

From Pilot to Portfolio

A small step can grow into a portfolio habit. This section shows how we can take a few repeatable components, prove the method on live projects, then spread capacity through more than one producer or a near site set up. The point is not a one-off trial. The point is a stable way of working that holds its shape across projects through clear roles, fixed interfaces, simple gates, and clean evidence. Where it helps the reader, we link to Chapter 5 for quality and Chapter 6 for time.

Capability

Small scale MMC can build habits that travel across projects. We can pick two or three component families, for example, façade cassettes, service risers, bathroom pods, then run the same playbook each time. We should fix the interface once, freeze the details, and write a short method statement that the factory and the site both follow. The cycle gives hard numbers on right first time, install hours per unit, defect rate at arrival, and days from order to factory slot. The quality lead keeps QITP and QITC visible so checks are the same on every project, and the record is complete for handover (see Chapter 5). If we can make a small part of the job repeatable, then we carry that repeatability forward.

Capacity

A distributed production base reduces fragility. We can qualify more than one producer against the same interface rules and the same evidence pack. If one source pauses, the others can increase volume without redesign. Where the pipeline clusters in one area, we can set up a flying factory near the site for a season, using rented space, mobile jigs, and trained crews. This shortens routes, reduces damage in transit, and keeps cranes productive. The logistics lead books lifts and protection as part of quality, and the planner maintains a joint factory and site calendar so work stays in step (see Chapter 6). Configuration control must be tight so the product remains the same, wherever it is made.

Commercial

Light frameworks turn learning into a habit. We can onboard smaller producers with clear acceptance criteria and evidence, then place call offs sized to demand. Payment follows defined outputs and clean records, not brand names (see Chapter 3). Where results are strong, we can expand scope or add a second source. Where results lag, we can stop without large sunk cost. A short dashboard across projects keeps everyone aligned, and lessons learned flow back into the interface rules and the method statements. Over time, this becomes a portfolio pattern rather than a one off.

Flying Factory Near Site

In 2015 Skanska UK leased a 12 month, 400 square metre unit on the Perth Trading Estate in Slough to support Phase 1 of the Battersea Power Station redevelopment, the Circus West apartments. The team used the space to build fully serviced utility cupboards for the flats. Each cupboard was a steel frame unit, about two point four metres tall and one point seven metres wide, carrying the electrical board, heating interface, ventilation unit, and connections for appliances. The near site location kept the route short and predictable, with finished units travelling along the A4 into London.

The set up was simple and disciplined. Floor jigs fixed the footprint and the connection points. A small test rig checked pressure, electrics, and controls before release. The crew worked as one shift with named leads for design, factory process, logistics, and quality. Four gates held the flow. A design freeze locked the interfaces. A first article inspection proved the method and the jig. A pre delivery check confirmed labels, protection, and paperwork. An arrival check on site confirmed identity, condition, and evidence. These gates sat inside a short plan with QITP and QITC so every check had an owner and a record (see Chapter 5). The planner kept a joint factory and site calendar so lift plans and install windows stayed in step (see Chapter 6).

Output and benefits were clear. The factory produced around sixteen cupboards each week toward a total run of more than five hundred for the phase. Working indoors reduced rework and first round defects. Site cranes were used more productively because units arrived booked, protected, and ready to lift. Labour hours on site fell because the services were already grouped, wired, and pressure tested. The near site approach avoided the fixed costs of a permanent plant while giving most of the gains of off-site work. The lesson for a portfolio is practical. We can rent modest space near a cluster of projects, move a simple jig and a trained crew in, and make repeatable units to a common standard. If a distant supplier pauses, the near site base can steady flow without redesign, provided we keep configuration control tight and the record whole.

Supplier Concentration

Modern methods depend heavily on specialised products manufactured under controlled conditions. The tooling, certification requirements, and fixed interfaces naturally limit the number of firms able to supply any given element. This clarity about who supplies what brings order to the process, but it also creates a narrow pipeline that can be fragile. As outlined in the introduction, the MMC supply chain operates on repeatable rules rather than one-off orders. Central to this is the systems integrator, often the main contractor or a specialist, who coordinates interfaces, manages factory bookings, and controls changes. When suppliers are few, the integrator must plan not just for cost, but also for resilience.

Supplier concentration has a direct impact on project timelines. Traditional construction materials usually have multiple alternate sources. If a shipment of blocks or plasterboard arrives late or damaged, sites can often find replacements quickly from other merchants. With MMC components like bathroom pods, panelised façades, or MEP racks, the situation is quite different. These require specific jigs, skilled crews, and proven fixes. Lead times often extend to weeks or months. A delay in one batch can halt wall construction, leave service risers incomplete, and stall fit-out behind the affected area.

When supplier concentration leads to failure, the impact on the programme can be severe. In April 2022, Caledonian Modular entered administration while holding contracts for the Department for Education's school building programme. This resulted in unfinished sites, closed buildings, and the urgent need to re-procure. This case clearly demonstrates why designs must accommodate alternate suppliers, contracts should include step-in rights for drawings and jigs, and thorough evidence must be held so that another producer can take over with minimal redesign.

Caledonian had contracts to deliver new-build schools using modern methods. When the company collapsed, several projects were left partially built. Subsequent reviews revealed structural issues at completed sites, and three schools were slated for demolition and rebuilding. The immediate consequence was delay for pupils and staff; then came a secondary wave of costs as the client replaced the supplier and ensured building safety. Financial flows also suffered as creditors, including the department, faced losses during the administration. In short, the failure of one supplier created risks not only for time and cost but also for the reputation of the entire programme.

For a portfolio approach, the lesson is clear and practical. We must maintain stable interfaces and keep a complete, accessible record so another supplier can step in smoothly if needed. Contract terms should reserve factory capacity, define clear quality gates, and include step-in rights over all relevant drawings, models, and test evidence. The client team should maintain a second qualified supplier on standby for low volumes to preserve a genuine option to switch. Quality must remain visible through QITP and QITC processes, with test reports, certificates, and photographs stored centrally. When the records are robust and complete, re-tendering, re-testing critical details, and restarting production is possible with minimal delay. When records are weak, recovery becomes a redesign effort rather than a straightforward handover.

This exposure can be managed without losing control. The design should hold interfaces stable so more than one qualified supplier can meet the same connection rules. The integrator should enforce a formal onboarding process for each supplier, which includes checks on capacity, quality systems, and financial health. Contract terms can include reserved factory time, step-in rights if a supplier fails, and fair price adjustment mechanisms tied to clear drivers. Keeping a second qualified source "warm" at low volume preserves choice. Factory acceptance tests and slot readiness reviews must be scheduled on the programme, with gates confirming that drawings, materials, jigs, labour, and inspections are ready before production begins. For long supply routes, a small buffer of finished units can be held off site so that minor delays do not stop site installation. If the regional supplier base is limited, combining a distant supplier with a near-site flying factory for simpler sub-assemblies can steady the workflow. All of these strategies rely on tight configuration control, ensuring the product remains consistent across locations and the records stay whole (see Chapters 5 and 6).

Balancing quality and flexibility in this context requires care. Working with fewer specialist suppliers can simplify assurance because processes become stable, inspectors understand the method well, and data flows are consistent. However, this comes with the risk of reduced leverage when performance slips. Clear upfront rules, and payment tied to evidence and complete deliveries rather than brand names, help manage this. QITP and QITC processes define what good looks like from material intake through to site arrival. Test reports, certificates, and photos are stored together so the quality story is complete at handover. Regular joint reviews compare metrics such as right-first-time rates, rework, and install hours across suppliers. When a supplier underperforms, changes follow a controlled route rather than ad hoc fixes. When a supplier excels, volume can grow without altering the interface. This approach keeps quality high while maintaining options. It transforms a concentrated supply market into a manageable system, one capable of withstanding shocks without rewriting the entire job.

Limited Product Choice in MMC: Customisation Within Constraint

Modern methods bring clear benefits in speed, consistency, and quality, but the range of product types and configurations can sometimes feel limited. Traditional construction allows teams to tailor nearly every element on site, responding flexibly to unique conditions. In contrast, MMC relies on pre-designed, standardised parts made to tight tolerances within controlled factory settings. This discipline ensures reliability but also sets firm boundaries. Tooling, jigs, and certification processes favour repetition, while one-off changes drive up unit costs, slow production lines, and trigger additional checks at every stage, from factory to logistics to installation. For this reason, we should approach customisation as a design challenge to be solved early, not as a last-minute problem on site.

Why Choice Feels Limited

Factories create value through stable interfaces, fixed grids, and repeatable details. Even a small change to a wall cassette or bathroom pod can mean the need for a new jig, a different lifting method, or fresh testing. Because these costs must be spread across volume, low demand variants weaken the business case. Certification adds another layer of complexity. Type tests, fire ratings, and product markings underpin many standard elements, and introducing variants often forces retesting and reapproval. Transport considerations also matter, panel size, weight, and rigidity must suit the transport route, cranes, and hoisting equipment. These factors explain why modular façades work well for high-volume housing blocks but struggle in conservation areas where materials and window rhythms must match existing streetscapes.

This tension appears inside the building plan as well. A pre-configured kitchen or service pod might clash with an irregular footprint. The team can either redraw the layout to accommodate the pod or omit it entirely, resorting to traditional on-site construction for that area. The right choice depends on how much the change disrupts the system. Staying within known parameters lets the factory maintain pace, keeps quality checks straightforward, and delivers site benefits. Pushing beyond those boundaries slows production and increases the assurance workload.

Working Within Constraint

Smart design within limits is possible. Many platforms now support what might be called mass customisation, the base product stays standard, but selected features can shift within fixed rules. Service entry points move within defined zones, finishes vary across a catalogue, and module lengths step in fixed increments. The design team should lead a short options study at the outset, using model-based rules to test layouts against real product constraints. The goal is to develop a scheme that fits the system rather than fights it. As Chapter 2 explained, stable interfaces make this possible, while Chapter 6 stresses the importance of an early design freeze to protect the schedule once choices are made.

Supplier engagement should start early. Inviting manufacturers to a defined gate allows confirmation of what can vary and what must remain fixed. That meeting should conclude with a simple playbook, covering grids, tolerances, service zones, lifting points, and protection rules.

This playbook then becomes part of the information model and contract. From this foundation, value engineering focuses on selecting and configuring within agreed rules, rather than late-stage brand switching. Chapter 7 highlights how this approach protects costs by maintaining volume on standard parts and avoiding expensive bespoke variants.

Digital tools help keep the team inside these rails. A product-rich model allows clash detection, installation rehearsals, and cost and carbon comparisons before orders are placed. It simulates how a brick slip façade panel fits onto a frame, how a pod negotiates a corner, or how a cassette clears a stair core. This isn't about extra polish, it's about avoiding late changes that disrupt factory rhythm and the quality record, connecting closely to the assurance model discussed in Chapter 5.

Hybrid Strategy

Some situations demand flexibility that MMC cannot provide, such as a listed façade requiring hand-laid brickwork, a complex atrium, or a unique staircase. For these cases, a hybrid strategy is best. Structural frames, service risers, and bathrooms can follow MMC, while heritage façades, staircases, or feature spaces use traditional trades. Interfaces between these two worlds must be clearly defined. The design lead should specify connection details, tolerances, and sequencing so the site doesn't become a battleground where systems clash.

This hybrid approach protects value by capturing repeatability where it matters, while preserving architectural expression where needed. The factory stays efficient, the site team keeps control, and the quality record remains intact. Decisions should rest on evidence – if a change triggers new jigs, new tests, or long lead times for a single feature, it may be better done traditionally. If it fits within the platform's standard ranges, it should remain in the factory.

Roles and Gates

Clear roles turn constraints into control. The design lead owns the interface rules and the design freeze. The supplier lead confirms what can vary and keeps the configurator current. The quality lead sets and monitors QITP and QITC so any variant is tested, labelled, and recorded as rigorously as the base product (see Chapter 5). The planner maintains a joint calendar ensuring long-lead decisions happen on time (see Chapter 6). The commercial lead locks outputs and evidence into contracts, so payment follows performance (see Chapter 3). With these roles and gates firmly in place, we can work confidently within constraints, preserve choice where it matters most, and avoid costly late changes.

MMC Operational Realities

Rhythm and Flow

Modern Methods of Construction (MMC) rely heavily on rhythm and flow once the design phase has been completed. The process moves away from abstract ideas to focus on the smooth and steady movement of components through factory production, logistics, inspections, and site preparation. Every stage depends on the others working in harmony, and clear responsibilities must be established to maintain this flow. The pace is no longer dictated by the on-site work but by upstream factory scheduling and production slots.

Lead Time and Design Freeze

Lead time behaves differently in MMC compared to traditional construction. In conventional projects, the site controls the schedule, with materials and trades adjusted to fit changing conditions. However, in MMC, the factory's production timetable governs the entire process. Missing a booked production slot can cause significant delays and disrupt the entire workflow. Before manufacturing begins, the design must be fully frozen since changes after this point are costly and complicated. Unlike site-based construction, where minor errors can be corrected on the fly, MMC requires designs to be fixed and interfaces clear. To avoid risks, production should only start when all components such as drawings, bill of materials, and inspection plans are complete. If anything is missing or incomplete, it is better to delay production than risk problems later. The chapter illustrates this concept with a bathroom pod example, highlighting the importance of managing design freeze and change orders carefully to protect the production schedule.

Procurement and Supply Chain

Procurement in MMC goes beyond simply ordering materials – it includes securing factory production time. Suppliers require stable and predictable demand to plan their workforce, tooling, and material orders efficiently. Framework agreements and volume contracts help suppliers prioritise orders, reducing risks and smoothing the flow of deliveries. Aligning freeze dates and consolidating smaller orders into larger batches further improves factory efficiency. Planning teams and procurement leads must work closely to coordinate schedules across different projects, using shared calendars and commercial tools introduced in earlier chapters.

Logistics and Site Delivery

Logistics is especially complex in MMC because large factory-made units such as pods or cassettes must arrive on site in precise order and at exact times. This requires pre-booked delivery slots coordinated with crane windows and clear space on site for offloading. Weather conditions and urban site constraints add further challenges, making tight delivery windows essential to avoid delays. The logistics plan must ensure the site is ready to receive units, including confirmed access routes, safety zones, crane availability, and concrete readiness. Offloading crews follow detailed scripts to keep the site organised, while nearby marshalling areas help manage limited laydown space. Planning for alternative weather days and realistic buffers in both time and space helps maintain a calm and predictable delivery flow.

Transport and Route Constraints

Transport limitations influence MMC design and logistics significantly. Large units must fit within legal road limits regarding width, height, weight, and turning radii, with transport permits and escort requirements adding complexity. These factors affect delivery timing and cost, often necessitating splitting large assemblies into smaller parts to fit within transport envelopes. Early route surveys and consultations with hauliers help confirm legal windows and identify risks. Coordination between logistics leads, hauliers, and site teams is vital to agree on arrival sequences and contingency plans. Buffer stock staged nearby supports continuity in case of minor delays, keeping the overall flow steady.

Site Preparation and Factory Readiness

On-site assembly depends on the factory delivering parts that fit precisely. Factory tolerances are tight, often within five millimetres, so site bases must be prepared to similar standards. If slabs are uneven or out of level, pods or panels will not fit correctly, leading to delays and increased costs. Achieving this "factory ready" status requires permanent survey points, laser scans, and templated set-outs confirmed by design and quality leads. Minor deviations can be adjusted with packers or grout, but significant errors require correction before delivery. The installation of the first unit is carefully observed and used to establish standards for lifting, fixing, bracing, sealing, and acceptance. After this initial "first set", repetition allows faster installation and more predictable crane days, speeding the overall programme and simplifying handover documentation.

Quality and Inspection

One of the strongest benefits of MMC is the improved quality resulting from controlled factory conditions. Factory environments enable consistent workmanship, trained operatives, and repeatable processes, which reduce variability compared to site construction. Quality inspections are planned stages integrated into the production process rather than reactive checks on site. First article inspections ensure the complete unit meets standards before volume production begins. Subsequent batches undergo testing and inspections following clear plans, with authority granted to halt production if defects are found. Quality control extends through transport and site arrival, where packaging, weather protection, and labelling are tightly controlled. On-site arrival checks confirm unit identity and condition before installation proceeds. Through careful information sharing and controlled change management, design, logistics, and site teams ensure the factory delivers exactly what the site expects, resulting in reliable quality that simplifies handover.

Certification and Compliance

Certification in MMC involves demonstrating both product and process compliance. While standard building regulations apply, factory-made elements require additional certification layers such as type tests for fire resistance, acoustics, and structural integrity. Factory production control evidence is also necessary. Early clarity about which components are covered by standard tests and which require project-specific inspections, especially at site junctions and penetrations, is essential. Roles are clearly defined: manufacturers take responsibility for product evidence; installers ensure installation conformity; principal designers manage compliance mapping, and clients oversee warranty strategies. Payment schedules are linked to evidence collection to ensure certification is built during the process rather than after completion. Simple matrices and clear gate criteria guide compliance from design freeze to installation, maintaining traceability and warranty validity throughout.

Environmental Performance

Environmental benefits from MMC are only realised when design and procurement focus on measurable data rather than slogans. Factories reduce waste through careful measurement of offcuts, controlled packaging, and working in weather-protected environments, resulting in fewer skips and spoil on site. Routine reporting on material yields, energy use, and landfill diversion sets clear environmental targets. Supplier and logistics teams support these efforts by managing returns and backhauls effectively. The quality lead reviews environmental performance alongside product quality data to ensure continuous improvement.

Compatibility and Interfaces

Compatibility is critical because factories build exactly what they are instructed, leaving no room for assumptions. Connection points, service clearances, and tolerance stacks must be validated early to guarantee parts fit correctly. Design leads publish interface rules and approved penetration lists. Digital models are used to detect clashes before production orders are placed, reducing the risk of costly errors. Early site readiness checks confirm levels and access to prevent delays or product returns. This careful attention to compatibility improves installation flow and supports environmental goals by avoiding unnecessary rework.

Process Readiness

Process readiness is just as important as geometric precision. Receiving crews need clear, concise instructions about lifting points, temporary bracing, torque settings, and sealant application. Utility connections such as power and water must meet specifications. Site managers lead arrival checks to verify unit identity, condition, and paperwork before lifts commence. Training at the start of production runs familiarises crews with fixings, tools, and sequences, often using sample frames. The first set installed under supervision becomes the standard for the rest, improving speed, predictability, and handover quality.

Standardisation and Controlled Variation

Standardisation adds value when parts are easy to select and difficult to misuse. A controlled library of option packs offers designers reliable choices with predefined weights, fixings, and lifting rules, enabling suppliers to quote consistently and site teams to become proficient with recurring details. Over time, this accelerates work and stabilises inspections. Controlled change management prevents deviations from standards, ensuring standardisation supports both efficiency and quality without limiting project-specific design needs.

Controlled variation balances uniformity with flexibility, allowing standard shells to accommodate different finishes or arrangements. This approach maintains character while preserving production line efficiency and quality documentation. Variants are clearly identified by design leads, reviewed as first articles, and paid for through the same inspection and certification processes as the base product. This system integrates environmental, compatibility, and process goals, providing measurable targets, clear responsibilities, and full traceability. The outcome is confident, efficient construction that fits correctly first time and delivers consistent quality.

Project Specific Suppliers in MMC

Project specific suppliers are companies commissioned to produce unique products tailored for a single project, such as a bespoke facade cassette, a modified pod, or a customised services module. These suppliers come into play when the project's requirements or context cannot be satisfied by standard systems or catalogue options. While this approach offers genuine freedom to meet particular needs, it also places significant demands on timing, coordination, and assurance. When used thoughtfully, project specific suppliers can expand the possibilities of modern methods of construction, but if engaged without careful control, they tend to introduce risks without delivering proportional benefits.

When We Should Use Them

Turning to project specific suppliers is appropriate when a standard platform cannot satisfy a clearly defined need. This might occur with a conservation frontage requiring sensitive detailing, a complex geometry that defies typical modular shapes, or a serviced core that must fit within an irregular plan. The decision to pursue a bespoke route should happen early in the design process, ideally before interface rules are fixed. The design lead plays a crucial role in setting a concise statement of need, defining the desired performance outcomes, and establishing the allowable ranges for size, weight, and access. Engaging the supplier at this stage invites collaboration on developing a solution that respects design for manufacture and assembly principles. This early involvement reduces the risk of designing a visually appealing product that cannot be efficiently produced within factory constraints.

Risks to Manage

The primary risks associated with project specific suppliers include fragility, delays, and a lack of transparency. A company strong in prototyping may struggle when required to scale up production volumes. Financial stability can be limited, and digital capabilities might lag behind, complicating model-based coordination essential to MMC workflows. Changes made late in the process can have a significant impact because the tooling, materials, and testing regimes are often bespoke and tailored to a single configuration. To mitigate these risks, a straightforward set of pre-award checks should be conducted. These checks assess capability against required tolerances and process controls, capacity and lead times with named crews, digital readiness for model exchange rather than simple sketches, and financial health supported by evidence. A clear plan for type testing, product marking, and certificate ownership is also necessary, linking closely with procedures outlined in earlier chapters. From a commercial perspective, staged payments tied to deliverable evidence, performance bonds when appropriate, and step-in rights over drawings, models, jigs, and data provide essential safeguards.

How We Make It Work

Successful collaboration with project specific suppliers requires treating them as an integrated part of the design team rather than a black box. There should be three critical gates in the process. The first gate is an interface freeze, which finalises grids, loads, tolerances, services, and lifting points with the supplier present to ensure mutual agreement. The second gate involves a digital handover that confirms the shared model, naming conventions, and revision control, supporting the programme's scheduling needs described in Chapter 6. The third gate is a first article inspection that verifies a complete unit, checking every fixing, seal, and label against an agreed condition. Quality control then proceeds through a concise plan, employing QITP and QITC to ensure checks at the factory bench and upon arrival mirror those of standard repeat products. Change control must be tightly managed, with any modification passing through a transparent technical change approval process that evaluates the impact on design, testing, and programme before work proceeds.

Sustainability and Evidence

Unlike standard suppliers who often provide environmental product declarations and detailed factory waste data, project specific makers may lack such documentation. It is important to incorporate environmental requirements explicitly into the supplier's scope of work. This should include material origin lists, energy use per unit, waste diversion rates, and a simple transport plan that accounts for the return of stillages or packaging. If full evidence is not available at contract award, there should be an agreed pathway for collecting this data during production. The essential point is transparency; bespoke solutions should never obscure their environmental impact.

Leaving a Legacy

One-off products should not result in one-off learning. The project close-out should include documenting tolerances, fixings, installation hours, and common issues in a concise report. The commercial lead can negotiate licences that allow drawings, jigs, and method statements to be shared across the programme. When a product performs well, it can be moved into a small options library with clear rules, transforming it from project specific into a systemised solution. Where performance falls short, the reasons should be recorded to avoid repetition. This approach ensures that bespoke work adds capability and value rather than noise or risk.

Flying Factories in MMC

A flying factory is a temporary manufacturing setup located close to the construction site. This approach is chosen when full off-site production doesn't suit the project's needs, yet traditional on-site labour cannot deliver the required speed or consistency. The concept is straightforward: a small, well-organised production line is brought near the project for a limited period, producing one or two repeatable items before being dismantled. This method maintains factory discipline while gaining the advantages of local delivery and avoiding long transport routes that add time and risk.

The practical benefits of flying factories are significant. Shorter transport distances reduce haulage costs, minimise damage in transit, and make delivery schedules more predictable. The layout of the flying factory is clean and organised, allowing for a smooth flow of materials from receipt to dispatch, with a single first article, one set of tools, and a consistent inspection plan. Since the setup relies on rented space and simple jigs, upfront costs remain modest compared to a permanent facility. Hiring locally brings social value and enables a faster response to day-to-day changes on site. The controls required are the same as for a fixed factory; an interface freeze, slot readiness, first article inspection, in-process checks, and a final arrival gate on site are all essential, connecting to the quality assurance processes outlined in Chapter 5. The overall schedule still matters because being close does not remove the need for a coordinated plan across factory and site, as described in Chapter 6.

Flying factories should not be seen as shortcuts. They require essential infrastructure such as power, welfare facilities, safe access, permits, and competent supervision. Quality must be demonstrated through documented records rather than assumed. If the product requires certification, the temporary line must follow the same processes, including third-party oversight where necessary. When treated properly, a flying factory becomes a valuable option in the MMC toolkit rather than a last resort.

Crossrail Case Study

Crossrail, now known as the Elizabeth line, needed thousands of highly precise tunnel lining segments for the twin running tunnels beneath London. Each ring is composed of precast concrete segments manufactured to tight tolerances. If a segment does not fit, the tunnel boring machine must stop. To ensure supply and control delivery, the eastern tunnels were served by a dedicated facility at Chatham Docks in Kent, while the western tunnels had a separate facility at Old Oak Common. Each acted like a flying factory, set up specifically for the programme, operated with full factory standards, and later stood down.

At Chatham Docks, the team installed moulds, curing systems, and gantry handling equipment, running continuous casting operations around the clock. A small laboratory monitored mix quality and strength. Each segment was tagged to trace its mould, batch, checks, and test results. At peak, the line produced more than two hundred segments daily, casting approximately one hundred and ten thousand segments for the eastern tunnels over the project duration. The process mirrored that of a permanent plant, with a single agreed first article, clear work instructions at every station, and production and quality leads empowered to stop the line if standards were not met.

Proximity to the site changed the risk profile significantly. Finished segments left Chatham by water, travelling up the Thames before short road journeys took them to the launch sites. A single barge could carry many complete rings, drastically reducing lorry journeys and exposure to motorway disruptions and curfews. The short, predictable transport routes supported booked deliveries to shaft locations and kept stockpiles manageable. This same logic applies to building projects where volume justifies a near-site production line, allowing cranes to remain productive as units arrive sequentially with appropriate quality proofs.

Assurance was built into the workflow. Each segment was weighed, dimensionally checked, strength tested, and visually inspected before release. Digital tags linked every unit to its mould, batch, and test results, enabling any quality issue to be contained within a known batch and traced back to its production station. The facility also created local jobs, with a core skilled crew training new hires to safely operate the line, thus building local capacity throughout the programme's life.

The Crossrail example demonstrates that flying factories succeed when integrated into the supply chain from the outset. The setup included the same roles and quality gates as a permanent plant, matched logistics to production, and maintained a robust audit trail. The outcome was secure supply, reliable delivery, and the quality needed to keep tunnel boring machines continuously operational.

Local Suppliers in MMC

Local sourcing is appealing because it reduces travel, supports communities, and appears beneficial for carbon emissions. However, in modern methods of construction, it only works when local capability matches proximity. The starting point is to ask what the local market can produce to factory standards, rather than what it might learn on the job. In simple terms, local should also mean ready.

The first hurdle is technical readiness. Modern MMC products require tight tolerances, consistent finishes, and exact interfaces. While many local firms excel at traditional craft and trades, they may lack the jigs, gauges, or quality control routines necessary for repeatable manufacture. Digital fluency is also critical. Most projects operate through shared digital models, meaning suppliers must be able to read, update, and revise models with correct naming and version control, as discussed in Chapter 6. Without this, coordination slows and the risk of errors grows.

Next is process discipline. Even if work isn't fully off-site, modern methods depend on design for manufacture and assembly, clear work instructions, and traceability. It is essential to verify that local suppliers can run a first article, capture batch data, and maintain records that feed into the golden thread of quality assurance (see Chapter 5). Scaling up is another challenge. A firm that performs well on one site may struggle to deliver consistently across multiple towns or projects. Financial strength also matters. A small balance sheet can turn a late payment into a production halt.

When the fit is right, the benefits are tangible. Shorter delivery routes reduce emissions and damage during transit. Local crews can visit sites quickly for surveys, trials, and problem-solving. Face-to-face collaboration improves trust and accelerates decision-making. Achieving this state requires deliberate supplier development rather than hope. Pairing experienced manufacturers with local firms, running onboarding workshops, sharing method statements, and setting up small near-site assembly trials help build confidence.

The commercial framework should reward proven delivery. Framework agreements provide pipeline visibility, encouraging firms to invest in tools and personnel. Initial call-offs can be modest, growing in size as evidence of capability accumulates. Contracts should link payment to clearly defined outputs and records, rather than brand names. Certification is a journey. If standards are not fully established, audits, templates, and mentoring can keep warranty routes open. When managed this way, local supply ceases to be a slogan and becomes a reliable, integrated part of the MMC system.

Sustainability and Carbon in MMC Supply Chains

Sustainability in modern methods starts upstream. While site teams can manage a clean working environment, the real environmental footprint is often determined long before materials leave the factory. To understand and control this, we need to look at the entire supply chain, from raw material extraction and manufacturing through to delivery, use, and what happens at the end of the product's life. This means measuring embodied carbon, making informed design choices, sourcing responsibly, and maintaining clear records as proof.

What We Measure

Embodied carbon refers to the carbon emissions involved in making, processing, and transporting materials before they reach the site. Environmental Product Declarations (EPDs) provide verified data at the product level, such as per square metre of panel or per fixing unit. Taking a whole life perspective, often through a life cycle assessment (LCA), combines the impacts of products, transport, maintenance, and end of life processes. Guidance from RICS and emerging national standards help set targets and create a common language. None of this works without accurate supplier data. Factories must report what each unit contains, how it was made, the distance travelled, and the waste generated. Modern methods help because assemblies repeat; modelling one panel accurately allows scaling over the full production run.

Design Choices

The carbon footprint is ultimately shaped by what we decide to build. Lightweight, material-efficient systems reduce emissions if they meet performance requirements for strength, fire resistance, sound, and thermal qualities. Panels can replace heavy blockwork, and volumetric units can substitute complex on-site builds while reducing waste. These benefits only hold if the product is well made and does not travel unnecessarily far. Chapter 2 explained how platforms and material categories influence these decisions, while Chapter 7 explores how cost and carbon can be balanced in value terms rather than price alone.

Sourcing

Location plays a vital role. Short transport routes reduce carbon emissions and damage during transit. Local supply can be strong when capability exists, while central plants may be superior where cleaner processes and better material yields apply. The best choice depends on the product, distance, and reliability of each supply chain. Procurement leads should evaluate both local and central options, backing decisions with data. Where local capacity is nearly ready, pairing experienced manufacturers with local firms can build capability.

Waste Control

Factories can transform waste management from guesswork into precise measurement. Offcuts are tracked and reused, packaging is managed carefully, and landfill diversion can be demonstrated. Setting simple targets for yield, waste, and energy per unit should be part of the project scope. These data feed into the golden thread of quality assurance with the same importance as product quality. If factories do not report openly, waste simply shifts from site to factory instead of reducing overall. The quality lead should review waste and energy figures at the same quality gates as product checks (see Chapter 5).

Transport

Transport contributes significantly to the total carbon footprint. Heavy units and long distances quickly increase emissions. We can reduce this load by designing products to fit within road transport limits, increasing packing density, and using near-site assembly where volume allows. Consolidated deliveries cut down on small van trips and idle time. A well-coordinated calendar ensures booked delivery windows that reduce waiting and repeat trips (see Chapter 6). Suppliers should report carbon emissions per load, showing the impact of weight, distance, and the number of deliveries.

Standards and Certifications

Sustainability claims require backing by recognised standards. These should be treated as design and procurement requirements, not just badges to add at the end. Whole-project schemes like BREEAM and LEED set targets and award credits for responsible sourcing and verified product data. For materials, FSC certifies timber from responsibly managed forests, and BES 6001 covers responsible sourcing for steel, cement, and others. UKCA marking confirms that products meet UK safety and performance rules. PAS 2080 provides a method to manage carbon for infrastructure projects, useful for programmes with multiple suppliers contributing to a shared outcome. Cradle to Cradle certification focuses on circularity and human health, which supports choosing components designed for reuse or recycling at end of life. EPDs and life cycle assessments provide the verified numbers. An EPD gives declared embodied carbon and other impacts for a defined unit, while an LCA aggregates these across the entire building or asset. Tender packs should request EPDs and specify tools for life cycle modelling so suppliers know how their data will be used. Repeatable assemblies make this practical, as one model can apply to many units.

A compliance plan should be written at design freeze. This plan names the project-level scheme, such as BREEAM Very Good or LEED Gold, and lists the product-level proofs accepted, for example, FSC for timber, BES 6001 for other materials, and UKCA where required. It also specifies the data expected from suppliers: EPDs, factory energy per unit, waste and diversion rates, and transport details like distance and number of deliveries. Ownership is assigned clearly: the design lead maps credits to evidence, the procurement lead incorporates requirements into contracts, the supplier lead provides EPDs and factory data, and the logistics lead reports transport mileage and delivery counts.

Traceability

Digital records turn sustainability claims into verifiable proof. Factory logs, QR codes, and batch tracking link each unit to its contents, tests, and transport route. On arrival, checks tie the product to its data, enabling credible reporting for clients and funders and speeding audits for schemes like BREEAM. This system also improves quality control because if a fault appears in a batch, the problem can be contained without guesswork.

Whole Life and Circularity

Durable assemblies built dry and to tight tolerances often last longer and require less maintenance. Planning for the end of life is equally important. Can units be removed and reused? Can parts be separated for recycling? Good systems avoid high-bond adhesives that prevent recovery. Where possible, products with clear routes for reuse or safe recycling should be selected. Responsible sourcing marks like FSC for timber and recognised material standards add confidence. Warranty, certification, and end-of-life planning should be integrated rather than treated as separate documents.

Putting It to Work

Carbon must be a design requirement, not just a report produced at the end. The design lead sets measurable outcomes at design freeze. Procurement selects suppliers based on proven data and reserves capacity with capable makers. The supplier lead provides product EPDs, factory energy use, and waste data. The logistics lead reports transport distance and deliveries. The quality lead collects and verifies evidence at each gate, ensuring a complete record at handover. When done this way, modern methods transform sustainability from a vague claim into a managed outcome with numbers the team can trust and choices visible to the client.

Logistics and Transport in MMC Supply Chains

Modern methods transform logistics from simple "trucks and forklifts" into a carefully choreographed process. Finished units arrive large, heavy, and depend on strict sequencing. One missed crane window or blocked road can halt a whole day's work. What follows explains how we plan the flow, prove the route, and keep lifts moving smoothly.

Context

In traditional construction, sites order what they need as they go and can often reshuffle trades if deliveries are delayed. In modern methods, the sequence is fixed months in advance. Drawings, lifting points, factory slots, and site layouts all rely on transport assumptions agreed early on. If these assumptions change late, the result isn't just a simple reschedule; it's lost time and costly rework. This is why logistics must join design at the start, as the time model in Chapter 6 highlights.

Planning

Just-in-time delivery works only when the site, haulier, and factory operate on the same timetable. Lifts should be planned as carefully as structural elements, with a clear calendar showing who moves what, where, and when. A readiness check the day before confirms slab strength, levels, exclusion zones, traffic controls, and access. If any element isn't ready, the delivery does not leave. Buffering helps – since urban sites rarely have space, a small marshalling point, temporary yard, or flying factory near the project can safely hold a few information-complete units. Think of it as a runway where flights land only when the strip is clear.

Routes

These aren't simple pallet drops. Units are wide, tall, and sometimes fragile. Routes must be surveyed in the actual vehicle, confirming bridge heights, lane widths, turns, curfews, weight limits, and escort requirements. Permissions can take weeks, so book early and keep a fallback route ready. When the product can't drive through a route, we adapt, splitting a cassette with a simple splice or sending a knock-down kit for near-site assembly. Nearby consolidation points reduce multiple drops and allow stillages and frames to return quickly so the gate doesn't get clogged.

Site Interface

Crane time is limited. Wall cassettes require mechanical lifts, pods need tower or mobile cranes, and some schemes use both. The lift plan sits alongside the programme, not beneath it. We must book windows considering weather and name an alternate day at the start. The lift supervisor owns the plan on-site, supported by banksmen, tag lines, and a simple script detailing stop points, set-down areas, and protection steps. When a flying factory feeds the job, gantry lifts move units to dollies, and routes from door to hoist are clearly marked and kept clear.

Risk Points

Transport can fail in many ways. Units may be damaged in transit, delayed by traffic, rerouted by police, or refused at a gate that's not ready. The cost isn't just the truck's running meter; it breaks the sequence. Follow-on trades can't progress, and the floor above remains open. We must plan responses, not rely on luck. The calendar includes small floats around critical lifts, marshalling points hold modest buffers, and teams maintain standby tasks so a lost lift doesn't become a lost shift. If a unit arrives damaged, the arrival check captures photos and moves the unit into a non-conformance process so cause and remedy are traceable (see Chapter 5).

Handover

Handovers must be seamless. The manufacturer passes identity and condition to the haulier, the haulier hands the manifest to the gate, and the gate hands control to the lift supervisor. No gaps allowed. QR tags and simple apps help, but only if used consistently. The logistics lead monitors flow live, confirms arrivals, and escalates issues while there's still time to act. After the day, data shows missed lifts and reasons, so future runs improve rather than repeat mistakes.

Roles

Fragmented control is the biggest risk. If the factory books the truck but the site books the crane, delays are inevitable. A single logistics lead should own the entire chain, from factory door to installed unit. The design lead confirms lifting points, fixings, and protections match the product. The haulier owns vehicles, timing, and permits. The site team owns readiness and lifts. Everyone follows one plan.

What Good Looks Like

Logistics joins at RIBA Stage 2 or 3, routes are proven before manufacture, lifts are booked with alternates, and readiness is checked the day before every drop. Units are tracked, handovers leave a clear trail, and one person oversees the entire process. Done this way, modern methods stop feeling brittle. The crane builds instead of waits, the road is a known constraint instead of a surprise, and the project moves with a steady rhythm the team can rely on.

Supply Chain Quality Monitoring

Quality in modern methods extends through the entire supply chain, not just on site. Much of the checking happens upstream with stage inspections in the factory, batch sign-offs, and type tests, ensuring units arrive finished and ready to lift. Making this flow visible is essential. The process starts with the factory proving one complete unit, followed by ongoing in-process checks. When the haulier takes custody, they verify the unit's identity and condition at the gate. On arrival, the site team carries out an arrival check and a brief acceptance before installation. This record travels with the product, maintaining the integrity of the golden thread. While Chapter 5 covers the standards and finish, this section focuses on the handovers.

Clear roles and ownership prevent gaps in responsibility. The manufacturer is responsible for product evidence, the logistics partner for custody and damage checks, the installer for correct fitting and safe fixing, and the principal designer for the compliance plan. Contracts (see Chapter 3) should link payment to evidence at each gate and specify who resolves faults and when. Digital tools such as QR tags, tablets, and photos aid in tracing each unit, minimizing disputes. Special care is needed for services, where interfaces should be proven using mock-ups and clear templates. Standardising pipe and duct zones speeds up checks, ensuring the site team is never the first to discover defects.

Chapter Summary

This chapter explained how an MMC supply chain functions as a connected system. It began by considering the external forces that shape outcomes, factors such as global sourcing, macroeconomic shocks, Brexit, pandemics, energy prices, climate events, labour availability, transport issues, digital risks, and insurance. The appropriate response is to plan for resilience, rather than just the lowest price, by fixing interfaces, using standardised and repeatable elements, and incorporating trials and simulations into the design phase using digital tools. This approach ensures decisions are validated before orders are placed.

The chapter then described how to operate the supply chain on a day-to-day basis. Lead time is earned at the factory, making early design freeze and slot readiness vital. Logistics must be treated as a timed operation involving routes, permits, crane availability, and scheduled arrivals, not simply a delivery. Flying factories, near-site, and temporary production lines can add capacity and reduce transportation miles when volume justifies their use. Local suppliers are effective when their capability matches proximity. Project-specific suppliers can extend the system's reach when treated as design partners with clear process gates, evidence requirements, and step-in rights.

Quality and sustainability are embedded throughout the workflow. Quality Inspection and Test Plans (QITP) and Quality Inspection and Test Certificates (QITC) ensure checks from material intake through to site arrival. The golden thread maintains a complete record. Carbon impact is considered through product choices and distance travelled, using Environmental Product Declarations (EPDs), life cycle assessments, factory yield data, and transport metrics to make informed decisions. Standardisation with controlled variation helps maintain pace while accommodating the project's context and brief.

Chapter 5

Product Quality Management

Introduction

Modern methods of construction (MMC) are often promoted as a guaranteed path to better quality. With factory precision, standardised processes, and controlled conditions, the assumption is that components built off site will naturally outperform those built traditionally on site. There is good reason for this confidence. MMC can reduce exposure to weather, allow for greater repetition, and apply more rigorous checks at each production stage. However, while the method provides potential, it does not guarantee outcomes.

High quality in MMC does not come automatically. It is not embedded in the module, the pod, or the panel simply because it was made in a factory. Instead, quality must be planned, managed, and verified throughout the process. Clear responsibilities, accurate design information, robust inspection routines, and well-managed interfaces are all essential. When one of these elements is missing, the impact can be far greater than in traditional construction. A single oversight may affect an entire batch of units. Similarly, if quality slips when components arrive on site, for example, due to poor storage or rushed installation, the benefits of the factory environment can quickly be lost.

This chapter explores how to understand and manage quality in MMC, not as a promise, but as a process. It begins with the factory setting, highlighting how controlled conditions can improve outcomes but also where they can fail, especially when design data is incomplete or inspection systems are too passive. It then explores accountability, asking who truly owns quality when products move through multiple hands before installation. These issues connect closely to the procurement and responsibility frameworks discussed in Chapter 3.

The chapter also examines what happens between the factory and the site. Transportation is often overlooked, yet it plays a critical role. Prefabricated elements can be damaged in transit, which introduces delay, additional cost, and repair challenges. Chapter 4's supply chain discussion is especially relevant here, as managing movement and handover between suppliers, logistics partners, and installers requires tight coordination.

We then turn to testing, certification, and accreditation (TCA). These processes help establish whether a product meets standards, but certification alone does not guarantee quality on site. The chapter explores how TCA systems are evolving in MMC and what gaps still need to be addressed to ensure compliance leads to real-world performance.

Risk management is another key focus. Poor quality affects more than aesthetics; it introduces risks to programme certainty, cost control, and long-term performance. Projects that fail to manage quality early often pay the price later through delay, re-work, or reputational damage. These time-based impacts tie directly into the issues raised in Chapter 6 on programme control.

DOI: 10.1201/9781003570431-5

Finally, the chapter addresses the challenge of re-work. In traditional construction, small errors can often be corrected on site with minor disruption. In MMC, the same errors may require factory returns, repeat inspections, or even replacement of entire modules. These consequences are not just expensive; they also affect project confidence, delivery certainty, and sustainability goals.

Throughout the chapter, the aim is to strike a practical balance. The goal is not to criticise MMC but to understand how to make it work better. By identifying common quality challenges, we also highlight what can be done to prevent them. This includes clearer contracts, stronger supply chain planning, better training, and improved use of digital tools. If these strategies are embedded early, MMC projects can deliver not only faster and cheaper, but better.

This chapter, then, is not just about identifying problems; it is about recognising the shared responsibility across clients, designers, suppliers, and contractors to ensure quality is not just expected, but achieved.

Off-site Factory-based Production: The Illusion of Control

Everyone talks about the factory as the solution for construction quality issues. In theory, an indoor factory environment offers controlled conditions, skilled labour, repeatable processes, fewer site variables, and no weather delays. These advantages do exist. However, assuming that moving construction off site automatically guarantees higher quality is a dangerous oversimplification. Factory production creates consistency of process, not necessarily consistency of outcome. If the design input is wrong, the output will be wrong – it's "garbage in, garbage out" on a massive scale. If the first module or panel is built with a mistake and no one catches it, the factory might repeat that error 100 times before realising it. At that point, you haven't just made one mistake, you've produced an entire batch of defective components. Worse, if this isn't noticed until units are delivered to site, or even installed, the impact is far greater and costlier to fix than a one-off error caught early. This is the illusion of control: the mere presence of a factory setting doesn't itself ensure quality outcomes. Without rigorous checks, a factory can simply mass-produce mistakes more efficiently.

On a traditional construction site, issues often get caught after assembling just a few units. A site manager might notice a misaligned wall or an incorrect installation, then immediately stop the work to correct course. In a factory, by contrast, work flows in assembly line sequence. Each team or station performs its task and passes the unit along to the next. Unless there are deliberate inspection gates in that workflow, an error will silently carry forward down the line. The system is only as good as the controls built into it. Modern methods of construction change the risk profile: they reduce many external risks but increase the importance of internal process control. A mistake on a site might affect a few units before correction; a mistake in a factory, unchecked, could affect dozens of units. The lesson is clear – off-site production can reduce variability, but it does not eliminate the need for vigilance. Quality must be actively managed, it won't happen automatically just because work is indoors in a repetitive process.

This is where the principles of inspection and test plans and quality certifications come into play. In traditional builds, a quality inspection and test plan maps out what needs to be checked, when, and by whom at each stage of construction. It forces ownership and accountability: each trade signs off their work, and each layer is inspected and tested before the next layer covers it. These quality processes were invented in manufacturing and engineering long before they were used on construction sites. The irony is that factory-based construction should inherently have better quality control, yet many off-site providers have not fully adopted those rigorous manufacturing QA practices. In the UK's off-site sector, for instance, some factories have been set up by traditional construction firms without deeply integrating engineering or lean manufacturing know-how. The result is that their quality control can be ineffectual – they are essentially attempting to build a construction product in a factory setting, rather than manufacturing a product for construction. To truly capitalise on the factory environment, the process must begin with an engineering and manufacturing mindset, not a construction site mindset. The focus needs to shift to input rather than just output. If production companies take this approach, quality isn't an afterthought; it is built into the process from the start. In other words, the factory can deliver better quality only if robust quality planning and checks are integrated into the factory workflow itself, just as rigorously as (or more than) on site.

Not only do we need a QITP for on-site assembly, we need a factory-aligned QITP as well, a detailed inspection and test plan for each module or batch produced off site. This plan should be built around the production stages in the factory, with clear "hold points" for inspection. For example, consider a typical sequence for a volumetric module:

- *Frame assembly*: Once the structural frame is welded or bolted together, check all dimensions, connections, and alignment against the design tolerances.
- *MEP installation*: After mechanical, electrical, and plumbing components are integrated, test that wiring and piping are correctly installed before closing up walls or ceilings.
- *Insulation and enclosure*: Before wallboards or cladding cover the insulation and vapour barriers, verify that all insulation is properly installed with no gaps or cold bridges, and that membranes or sealants are continuous.
- *Interior finishes*: Before the module is fully sealed, inspect that drywall, flooring, fittings, and fixtures are correctly installed and meet quality standards.
- *Doors, windows, and sealing*: Check that doors and windows are fitted square and plumb, operate smoothly, and are properly sealed against weather ingress.
- *Final clean and dispatch prep*: Conduct a final visual inspection after cleaning, ensuring no incidental damage occurred and that the module is ready for transport with protection and QA tags in place.

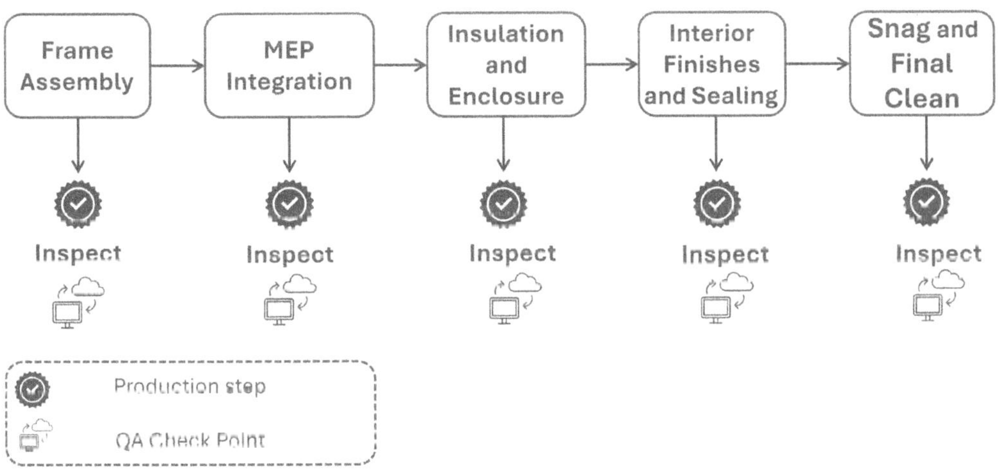

Figure 5.1 MMC QA Cycle

At each of these points, the station becomes a QA checkpoint with a documented sign-off. A good factory will have a checklist for each station, requiring photos or digital evidence to be recorded. The goal is that no module moves to the next stage unless the previous stage's quality checklist is completed and any issues rectified.

In practice, many established prefab factories do implement such in-process inspections – after all, they must adhere to building codes and often have to satisfy external certifiers. Reputable factories will have visible quality controls: signage about checks, station checklists, and dedicated QA inspectors or leads. Some use barcodes or QR codes on each module to link to its digital inspection records, others use tablet-based forms that site managers or workers sign off at every step. A robust off-site QA programme often includes measures like in-process dimensional checks with calibrated tools, electrical continuity and plumbing pressure tests before walls are closed, structural bolt torque inspections, and visual finish reviews under good lighting. In short, a well-run facility might have multiple quality control points: from verifying incoming materials and components all the way through to a final 100 percent module inspection before it leaves the factory. These controls, when properly implemented, provide evidence that each unit meets the specifications and code requirements – giving confidence to customers and accreditation bodies alike.

The problem is that not all factories operate at this high standard. Some newer or less-sophisticated off-site factories fall into the trap of assuming that because work is done indoors on a production line, quality must automatically be happening. It isn't. Repetition without feedback is just mass production – it's not quality assurance. If there's no feedback loop, you're simply mass-producing potential defects. And even with some controls in place, there are limits and challenges unique to factory building.

One such challenge is the tension between throughput pressure and quality. Factory workers are often under strict performance metrics. If something in a module does not align or fit correctly and would normally require extra time to fix, there is intense pressure to push through and keep the line moving. Pausing the production line is costly. This can lead to corners being cut or issues being ignored until later. Without a strong quality culture, throughput targets can trump craftsmanship.

Late design changes or unclear information also pose a risk. If design changes come in late, or if coordination drawings are unclear, the factory floor team might not have the capacity or authority to stop production and seek clarification. Their default may be to keep building according to the last information they have, even if it results in known issues. A better approach is to involve the factory at early design stages and lock the digital model before production starts.

The human factor remains critical. Many modular factories still rely heavily on manual labour and traditional construction skills. If workers are not trained in manufacturing quality practices, mistakes will happen. Without an embedded quality culture, even written procedures can be bypassed. If an assembly technician misses a bolt or applies the wrong sealant and there is no secondary inspection to catch it, that unit moves down the line with the defect hidden.

There is also the issue of lack of visibility into completed assemblies. Once a module or panel is enclosed, it becomes difficult to inspect what is behind the finishes. You must trust that everything done in earlier stations was correct. This puts a premium on process reliability and documentation. Thorough sign-offs before closure of walls, including photographic evidence of the hidden work, create an "x-ray" record for later reference.

Finally, certification overreliance can be problematic. Factories often point to certifications as proof of quality. These accreditations indicate that a quality system exists and that the design and process have been reviewed. They do not guarantee that every single unit is flawless.

Systems only work if they are rigorously implemented and continuously improved via feedback. Complacency after accreditation can be a risk.

Ultimately, the real question is not "is factory production better than site?" but rather "what monitoring and control systems are in place to ensure the factory delivers what it promises?" A well-run factory with proper quality processes can achieve superior quality and consistency – delivering on the promise of off-site construction. But that outcome is far from automatic. It requires meticulous planning and execution of quality assurance throughout the project lifecycle.

The path to true quality control in off-site production starts even before the first module is built. It begins at the design stage. The design and engineering team must have a deep understanding of the off-site production process and design for manufacturing and assembly. This means aligning the building design to the standard dimensions, tolerances, and process capabilities of the factory. Critical dimensions should align with the manufacturing line's constraints – if the factory's standard panel is slightly narrower than the architect's preferred module for aesthetic reasons, adjust the design to match the standard and avoid a custom setup. Often, designers resist such changes, but these small tweaks can make the difference between an efficient, error-free production and one riddled with bespoke adjustments and higher risk of mistakes. The design should be locked in before production starts, with all coordination of services resolved. Once production starts, there is little room for redesign.

Next comes the question of traceability and data capture during production. Do we know for each panel or module exactly which batch of materials it came from, which crew assembled it, and when? Can we trace a specific unit's journey: who performed each critical task, what inspections were done, where the unit was stored, when it was shipped, and when it was installed on site? If the answer is no to any of these, then we are not fully in control of quality.

This is where digital tools and a digital quality inspection and test plan help. The factory's stage gates are tied into the BIM model of the building. Each module in the model can be linked to its inspection record. When an inspector checks the MEP rough-in of a module and finds an issue, that gets logged in a database. When it is corrected and signed off, that update is recorded. The quality inspection and test certificate become the product passport for each module. It contains proof of all inspections done, any defects found and rectified, test results, and sign-offs by responsible parties. This travels with the module from the factory floor to the transport hauler, to the site assembly, and eventually is handed over to the client or asset manager as part of the building's documentation. If something ever goes wrong post-handover, there is no need to guess what was done or who is at fault – you can consult the module's QA record.

Implementing such comprehensive monitoring is not about bureaucratic paperwork; it is about risk management and accountability. Traceability, compliance, and documentation protect all stakeholders. For the manufacturer, it is a defence mechanism. For the client and insurer, it is peace of mind that the building was built as advertised, with verifiable checks. In sectors like automotive or aerospace, every critical component comes with a quality log and serial number; buildings assembled from factory-made components should strive for a similar level of rigour.

The factory is a powerful tool that creates the conditions for higher quality, controlled environment, repeatable tasks, and so on, but those conditions only yield better outcomes if actively managed with robust quality processes. Off-site production can significantly reduce certain risks, but it does not eliminate risk. In some cases, it can even create a new category of risk: a quiet, hidden, repetitive failure that only becomes visible after many units are produced or after installation. The antidote to that risk is a proactive quality regime – bringing the mindset of "build it right the first time, every time" into the factory, and not assuming that an indoor address alone solves quality problems.

Factories that fully embrace detailed QITPs, thorough in-factory inspections, and maintain complete QITC records, especially when supported by digital tools for traceability, not only protect themselves from costly mistakes, they also enhance trust across the supply chain. For clients, investors, and insurers, this approach turns an otherwise sceptical question mark into a reliable product. It is the difference between just mass-producing units and manufacturing a guaranteed product. The promise of off site will be realised when factories deliver on quality consistently, and can prove it with data. In the end, quality does not happen by accident, even in a factory, it happens by design, discipline, and diligent oversight at every step of the process.

Responsibility and Accountability: Who Owns Quality?

This is the question that surfaces when something fails and time is already lost. A panel is out of tolerance. A pod leaks. A service riser will not connect. Fingers point in every direction. The factory blames installation. The site blames transport damage. The designer blames a late change. The supplier blames unclear drawings. Meanwhile the client wants a single answer and a recovery plan.

In MMC, this confusion grows because more hands touch the product before it reaches the building. Factory teams, sub-assemblers, logistics providers, certifiers, installers, and inspectors all play a part. Somewhere in that chain, someone was meant to check, approve, or stop the work. The only way to avoid the blame game is to decide ownership before the first drawing is issued and before the first batch is made.

Start by naming one accountable party for quality at each stage. Do not assign an organisation. Assign a person with a job title and a phone number. In the design phase, the design lead owns coordination and tolerance. In procurement, the principal contractor owns the quality requirements in contracts and purchase orders. In manufacture, the MMC supplier's quality manager owns the factory inspection and test plan and the release certificate for every unit. In logistics, the haulier owns protection, load security, and damage reporting on receipt and on delivery. On site, the installation supervisor owns the site readiness check and the sign-off of every lift and fix. The client team or clerk of works owns acceptance against the specification. These names should appear on a simple responsibility matrix that sits in the common data environment and is updated when staff change. If a task does not have a named owner, you have a risk.

Next, connect responsibility to clear gateways. Before manufacture, a design freeze notice confirms what is fixed and who has approved it. Before dispatch, a factory release certificate confirms the unit meets the inspection and test plan and is fit to travel. Before installation, a site readiness certificate confirms the receiving structure, fixings, services, and access are ready. After installation, an interface inspection confirms that the unit is connected and protected. Each gateway needs evidence in the form of checklists, photos, and measurements. Each gateway needs a rule that says no pass, no progress. This is how responsibility becomes visible and enforceable rather than assumed.

Define who owns an error and how the clock runs. If a module arrives damaged, the haulier reports it at the gate and the factory quality manager is notified at once. If a lift is cancelled because the frame is out of tolerance, the principal contractor records it and the frame subcontractor is notified with a time limit for correction. If a service opening in a panel does not match the model, the MMC supplier is notified and a decision is made the same day on repair, replace, or workaround. A short non-conformance procedure with a single form and a single tracker prevents drift. It should state who raises, who approves, who fixes, and when. It should also state who pays, using the contract rules agreed in Chapter 3. When people know in advance how issues will be handled, they act faster and argue less.

Align responsibility with the quality records. The factory inspection plan and the site inspection plan must join up. The quality certificate for each unit should carry a unique identifier that links to the model, the drawings, the batch of materials, and the inspection photos. When the unit lands on site, the installer scans the code and the record continues with delivery date, lift location, fixings used, and interface checks. This is the golden thread that supports the product quality work in this chapter and the time control in Chapter 6. It turns opinion into evidence. It also helps the client and the insurer understand what was built and how it was checked.

Be clear about statutory roles and duty-holders. The design team is responsible for making the design buildable and safe. The manufacturer is responsible for building to that design and declaring conformity. The principal contractor is responsible for coordinating trades, controlling the site, and accepting work only when it meets the specification. Building control and third-party certifiers check compliance, but they do not own quality on the project's behalf. If the text of the contract suggests otherwise, fix it. Contracts should grant the client rights to inspect in the factory, set hold points, and require evidence before payment, while also protecting the manufacturer with fair rules on change and delay.

Traditional Quality Responsibility: Why MMC Needs More

Who owns quality? In theory, everyone. In practice, that often means no one. In MMC, we need clearly defined roles with real authority and a simple chain of custody, from drawing to delivery, installation, and acceptance. When a panel is out of tolerance or a pod leaks, we don't want finger-pointing. We want to know exactly who signed off on what, when they signed it, and what happened next.

Start with a positive truth: factory production can lift quality when managed well. Work happens in fixed locations. Tasks repeat. People become specialists in narrow activities. Monitoring is easier than on a live site. Cameras, time lapse, and CCTV can cover each station. Tablets capture photos at every hold point. Barcodes or QR tags link each unit to its records. This is a strong foundation. This section is not a criticism of factories; it's about making sure we use their strengths by putting responsibility where it belongs and making evidence routine.

Name the owners early. The factory quality lead owns the inspection and test plan plus the release certificate for every unit. The logistics lead owns protection, load security, and damage checks at loading and receipt. The installation supervisor owns site readiness and interface checks after each lift. The client-side reviewer, often the clerk of works or employer's agent, owns acceptance against the specification. Put those names, job titles, and phone numbers into a responsibility matrix stored in the common data environment, updated when people change. If a task has no named owner, that's a risk.

Link ownership to gateways. Before manufacture, a design freeze notice signed by the design lead, manufacturer, principal contractor, and client confirms what's fixed. Before dispatch, a factory release certificate with photos of hidden work and test results is required. At delivery, a receipt inspection with photos of the unit on the vehicle and after unloading takes place. Before installation, a site readiness certificate confirms the structure, fixings, services, and access are ready. After installation, an interface inspection records alignment, connections, seals, and any temporary protection. The rule is simple: no pass, no progress. These gateways protect time as well as quality (see Chapter 6).

Give the site team the right to look inside the factory. If the programme and contract allow, run periodic in-factory inspections by the principal contractor or client representative. Request photographic evidence of each manufacturing stage as a deliverable. Because work happens in fixed locations, factories can install cameras at each station and keep a visual log at low cost. This builds trust, supports the golden thread, and reduces disputes later.

Make the chain of custody explicit. The factory signs the unit over to the haulier with a condition report and photos. The haulier signs it over to the site with the same. The site logs condition and raises exceptions at the gate. If the unit is stored, the site agrees on protection, inspection frequency, and responsibility. These simple steps stop common arguments about when damage occurred.

Connect this to procurement, as Chapter 3 outlines. When the client buys pods from one source, panels from another, and installation from a third, the coordination risk lies with the client unless contracts move that risk downstream. Design and build can centralise responsibility, but only if contracts clearly define inspection rights, hold points, payment on evidence, and who pays for re-work. Purchase orders should not just state what's supplied; they should describe how quality will be proven and what happens if items are rejected.

Use a short and clear non-conformance procedure. One form. One tracker. It should state who raises the issue, who approves the fix, who performs the work, how long it will take, and who pays. It should link to the unit ID and the model. Close the loop with a photo and a signature. This keeps problems small and quick to resolve, while creating data for learning and for insurers.

Explain how this differs from traditional site-based responsibility. On a conventional build, the principal contractor oversees, subcontractors self-certify, and a clerk of works verifies. Because work is visible and trades overlap on site, mistakes are often spotted quickly and small fixes happen in real time. In MMC, that overlap is reduced. A unit may pass through several hands before anyone on site sees it. If one link skips a check, the error can be built in rather than built out. The answer is not to abandon MMC. The answer is to use its strengths. Factory production lends itself to tighter monitoring, photographic records, and repeatable checks. Workers can specialise and deliver consistent quality. That's a benefit, but it needs a clear line of responsibility to unlock it.

Bring the evidence into one place. The factory inspection plan and the site quality plan must join up. Every unit should carry a unique identifier linking drawings, model location, batch records, and inspection photos. When the unit arrives on site, the installer scans the code and the record continues with delivery date, lift location, fixings used, and interface checks. This is the golden thread in action, supporting both the quality goals of this chapter and the time control aims of Chapter 6.

Use an example to show how this works: a bathroom pod arrives with a cracked basin. Under the responsibility matrix, the delivery team photographs the crack at the gate and logs it against the unit ID. The site pauses the lift for that unit only. The haulier notifies the factory quality lead the same day. The contract sets the rules for who pays and whether repair or replacement is needed. If replacement is required, the factory books a new slot and the site reschedules lifts to keep the crane productive. No one argues where the basin was damaged because the chain of custody tells the story.

End with culture. Responsibility only works when leaders model it. Factory leads release units with complete records. Site managers refuse to install without readiness checks. Clients attend acceptance reviews and ask to see the evidence. When quality ownership is real, people act before problems grow. The tone is positive: factory production is good when done right. It allows more monitoring, more specialisation, and more control than a live site. MMC just needs a little more clarity on who owns what and better use of simple digital tools. With that in place, quality becomes routine rather than a debate.

Table 5.1 MMC Quality Responsibility Framework

Activity	*Responsible*	*Accountable*	*Consulted*	*Informed*
Design coordination sign, off	Lead designer	Client	Factory, contractor	Quality manager
Factory production stage checks	Factory QA officer	Factory construction manager	Client QA	Design coordinator
Pre, dispatch visual and dimensional QA	Dispatch QA lead	Factory construction manager	Logistics partner	Installer
Transport damage check (arrival)	Site logistics lead	Site manager	Crane supervisor	Client rep
Installation QA	Install subcontractor	Site QA lead	Clerk of works	Asset management team
Final pre, handover quality review	Client/site QA lead	Client	Principal contractor	Certifier/insurer

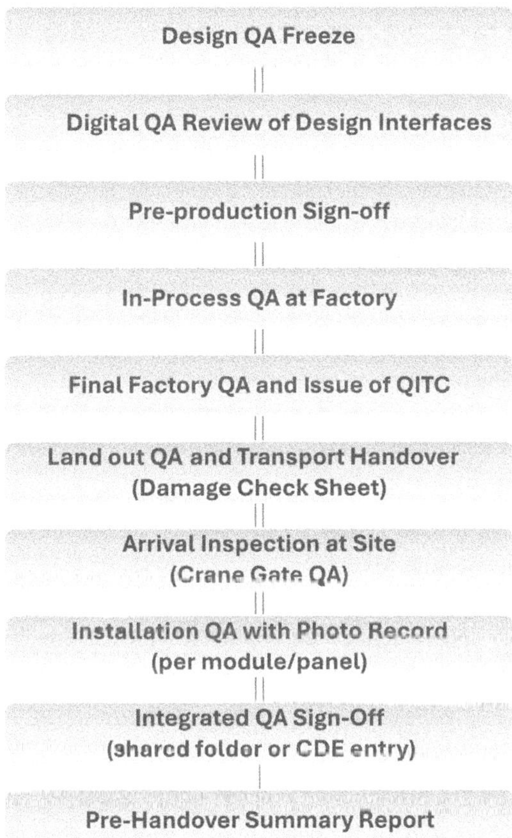

Figure 5.2 MMC Quality Workflow Map

Table 5.2 Key QA Documents

Stage	*Key document*	*Owner*
Design coordination	QA checklist, clash clearance log	Lead designer
Pre-factory production	Factory QITP, design freeze declaration	Factory QA lead
Factory stage inspections	Factory stage sign, off sheets, photo logs	QA officer
Pre-dispatch	QITC certificate, fit and finish checklist	Factory manager
Transport	Delivery condition sheet, sealant check	Logistics supervisor
Site arrival	Unwrap inspection log, lift readiness signoff	Site logistics
Install	Installation QA report, interface match sheet	Install subcontractor
Pre-handover	Combined QA summary, outstanding items tracker	Client QA lead

Key templates for MMC QA

These are not extra paperwork, they replace scattered spreadsheets, photos on someone's phone, emails that get lost. Every one of these forms should be pre, created and integrated into a code or shared dashboard. To form golden thread of information which will support building safety regulations and gives the client and insurer confidence.

Minimum fields every template should include

- Project, building, and unit ID
- Date, time, location or station
- What was inspected and against which standard or drawing revision
- Pass or fail outcome, measurements, photos, and comments
- Name and role of the person who checked and who approved
- If failed, NCR number, owner, action, due date, and close evidence

The following section is the bridge from policy to site practice. It lists the controls, the actions, the evidence and the named owners. Things can be set right from day one by using it to set weekly routines, book gateways, and link payment to proof in the CDE. This can be incorporated into inductions and programmes so quality becomes daily work everywhere.

How to Make this Work on Real Projects

For this framework to deliver real value, it cannot just live in a slide pack. It must be built into the project from day one. That means turning principles into clear deliverables and everyday routines, so responsibility and evidence aren't optional, they become how the job runs.

Create the QA Responsibility Matrix before Tender

Don't wait until after contract award. By then, bidders have already priced assumptions and gaps are baked in. Include a full QA responsibility matrix in the employer's requirements. List every touchpoint: design reviews, factory stage gates, delivery checks, installation verification. Name every document each party must produce and when. Include actual templates and controls. This links directly to Chapter 3 on procurement and ensures quality is priced in, not stripped out later.

Table 5.3 QA Control and Actions Matrix

No.	*Control*	*Key actions*	*Evidence*	*Owners*
1	Design for manufacture	DfMA brief Stage 2	Signed DfMA brief	Design lead
		Interface register Stage 3	Interface register	MMC supplier design manager
		Freeze geometry and risers by Stage 4	Tolerance schedule	Principal contractor design manager
			Design freeze notice	
2	Design sign-off by all	Product design sign off	Stamped PDFs in CDE with dates and names	Designer
		Production drawing sign off		Manufacturer
		Buildable and installable checks		Installer
				Client rep
3	Factory QITP in the line	QITP per product	Checklists with photos	Factory QA lead
		First off approval	First off report	Station supervisors
		Daily quality huddles	Calibrated tool logs	
4	Photo evidence of hidden work	Photos before closure	Time stamped images in CDE	Factory QA lead
		Photos before dispatch	Linked to unit ID and model	Install supervisor
		Barcode or QR per unit		
5	Site eyes in the factory	Monthly walk throughs	Visit reports with photos and actions	Principal contractor QA manager
		Remote video spot checks		Client-side reviewer
6	Chain of custody	Load out QA with photos	Signed handover forms with photos at each step	Factory dispatch lead
		Transport handover with photos		Haulier
		Gate check at site		Site logistics manager
7	Site readiness	Check structure, fixings, services, access	Site readiness certificate	Install supervisor
		Confirm lift plan and exclusion zone	Lift permit in CDE	PC construction manager
8	Installation QA	Check alignment, fixings, seals, hook ups	Install checklist with photos	Install supervisor
		Photo record per unit	NCR entries	Clerk of works
		Log deviations as NCRs		
9	Integrated sign off and pack	Integrated QA sign off in CDE	QITC linked	PC QA manager
		Pre handover summary pack	Install photos	Client-side reviewer
			Test results	
			Defect closures	

(*Continued*)

Table 5.3 (Continued)

No.	*Control*	*Key actions*	*Evidence*	*Owners*
10	Short NCR loop	One page NCR	NCR log with owner, fix, date, photos	Issue raiser
		Single tracker		Contract rules set payer
		Same day triage		
		Rules for repair or replace		
11	Monitor and learn	Track first time pass	Weekly KPI dashboard	QA leads factory and site
		Defects per unit	Actions closed next week	Project manager
		Inspection cycle time		
		Percent plan complete		

Assign Named People, not Generic Roles

Quality fails when nobody owns decisions. For each QA point, name a person, include their job title, phone number, and email. Add a deputy for cover. Publish this contact list in the common data environment (CDE) and keep it current. MMC is time-sensitive (see Chapter 6). When a unit arrives damaged or a lift point is missing, decisions need to happen in minutes, not days. Named people make that possible.

For each QA point record three contacts:

- Responsible person who performs the check
- Reporting contact who is informed if the check fails
- Fallback contact who covers if the others are unavailable

Use One Shared QA Tracker

Scattered spreadsheets and photos on phones waste time and erode trust. Use a single QA tracker as the source of truth. This could be Excel on a shared drive or a platform like Asite, ACC, Field-View, Zutec, or Procore. The key is that all inspections and photos are logged there, every item has a status and timestamp, and sign-off history is visible. Link the tracker to the CDE and to unit QR codes, so records open instantly with a site scan. Progress and payments should follow the evidence in this tracker.

Train Every Subcontractor on Their QA Part

Factory teams usually follow standard cycles; site teams work in changing conditions. Don't assume a shared language. Run a QA induction for every subcontractor and the factory team. Show the full flow from design to handover. Explain where their work fits. Walk through the exact forms they'll complete. Show examples of acceptable evidence, clear photos, dimensioned sketches, signed checklists with date and time.

Early adopters will win work. As MMC becomes mainstream, firms that train first, learn the process, and maintain clean records will be in demand. Make this a stated benefit when briefing the supply chain.

Put Quality Milestones into the Programme

Checks get skipped when time isn't allowed. Build QA gates into the master programme and micro-programmes. Give each gate real duration, clear predecessors and successors, and explicit float for fixes. List these as named activities on the Gantt chart tied to physical work, not as vague QA bars.

For example:

- Factory first-off approval before batch production
- Pre-dispatch hold point and issue of QITC
- Site readiness check before any lift
- Interface inspection after each install
- Cluster testing and sign-off before closing walls

Tie These Gates to Payments

No evidence, no progress claim; no sign-off, no payment. When quality is part of commercial control, it gets done (see Chapter 7).

Keep the Non-conformance Loop Short

Use one simple non-conformance form and a single tracker. Aim for same-day triage. Record who raised the issue, who owns the fix, the action, due date, and a close-out photo. Set clear rules for repair, replace, or accept with concession. Agree who pays in line with the contract. Short cycles keep problems small and protect the programme.

Prove the Product Before You Scale

Run a first-off sample at the factory and a site mock-up of the interface. Capture lessons learned and update the QITP and drawings before full production. Use this as a training aid for install teams and as a reference for acceptance. One good sample removes much debate later.

Put Site Eyes in the Factory

Schedule regular factory visits by the principal contractor and client-side reviewer. Fixed stations make monitoring simple. Ask for photographic evidence at each stage and make it a deliverable. If travel is difficult, use planned video walkthroughs. This strengthens the golden thread and chain of custody.

Manage the Chain of Custody

Quality suffers when damage isn't owned. Make handovers formal. At load-out, the factory records condition with photos and a damage check sheet. The haulier signs this and repeats the check at site. The site conducts an unwrap inspection at the gate and logs any issues before storage or lift. If units are stored, agree on protection and inspection routines and who is responsible.

Run Daily and Weekly Rhythms

Make quality a habit, not a reaction:

- Daily ten-minute quality huddles at the line and on site to review yesterday's issues and today's gates
- Weekly QA reviews that close actions, update the tracker, and clear blockers
- A visible dashboard in the site office and factory showing first-time pass rate, open NCRs, and overdue sign-offs

Build the Handover Pack as You Go

Don't leave documentation to the end. Each unit should build its pack as it moves through stages: design approvals, factory checks, QITC, delivery photos, install checks, test results, and close-out records. At pre-handover, produce a simple summary linking to the full evidence set. This is the golden thread supporting building safety regulations and giving the client confidence that what was promised was built.

Keep Roles, Documents, and Time Aligned

Real gains come when people, paperwork, and programme work together. The responsibility matrix tells us who acts. The templates provide the forms. The programme dictates the timing. When these three align, quality becomes routine and time becomes predictable. That is the essence of MMC, and how this chapter connects to Chapter 4 on the supply chain, Chapter 6 on time, and to Chapter 7 on cost.

The Cultural Gap: Ownership is More Than a Signature

MMC brings structure, technology, and process, but none of that matters if the people using it do not feel responsible. Quality is not achieved through a checklist alone. A culture where signatures are added without thought will not deliver what the system promises. Ownership is more than a form – it is an attitude that runs through the team.

The projects that achieve the best outcomes are those where teams understand why a check exists, not just that it must be done. When a factory operative knows how a missed seal will affect a bathroom pod weeks later on site, they are more likely to raise a concern immediately. When a site installer recognises how tolerances affect follow-on trades, they are more inclined to flag a misalignment instead of pushing it through. Reporting an issue early is not a failure, it is a strength, and it saves cost and time.

This is where project-wide QA inductions add real value. Bringing together factory teams, site crews, logistics leads, and design coordinators allows everyone to see the full picture. The exercise should not just be a briefing but a walkthrough of the end-to-end process, showing where each person's input connects with the next stage. A delivery driver who understands the inspection points will take greater care with handling. A site foreman who has seen the production process will better appreciate what can and cannot be changed after delivery.

The aim is to move from a culture where quality is enforced to one where it is expected. In MMC, the biggest gains come when everyone trusts the process because they helped shape it. That trust transforms quality from a compliance exercise into a shared mindset.

Responsibility Is the Infrastructure Behind Quality

Technology can help us manage quality. BIM models can visualise clashes, digital dashboards can track inspections, and certified factories can demonstrate standards. Yet none of these guarantee success unless responsibility is clearly owned and traceable. In MMC, where the margin for error is narrower, this becomes even more critical.

A strong quality framework is not about adding layers of control for their own sake. It is about giving teams the confidence to deliver without hesitation. When everyone knows who is responsible, issues are picked up quickly, decisions are made without delay, and the programme keeps moving. The clearer the roles, the smoother the workflow, and the stronger the outcome.

This means defining ownership before production even begins. Every sign-off should be linked to a named individual, not just a company or role. Each inspection should show who carried it out, when, and what evidence was recorded. This traceability is not bureaucracy; it is the infrastructure that allows MMC to deliver at pace. Without it, gaps appear, and in MMC, there is less tolerance for gaps than in traditional builds.

Quality is not a single task or a final inspection; it is an agreement shared across the team. It is the understanding that delivery only works when each person plays their part and hands it on in good condition to the next. In MMC, that agreement must be in place long before the first panel leaves the factory. When it is, projects benefit from smoother installation, fewer disputes, and greater confidence from clients and insurers alike.

Transportation Impact

This is often the missing middle of MMC quality. We checked the unit in the factory, and it passed every inspection. Then it arrives on site scratched, chipped, misaligned, or distorted, and no one can explain why. That is the risk of transport in MMC. It is the invisible stage where quality can quietly unravel. No alarms sound. No errors are logged. The unit simply changes condition between dispatch and arrival, and unless that change is tracked, the project inherits a hidden problem.

When that unit fails on site, everyone looks around for someone to blame. The factory insists it was fine when it left. The haulier says it was handled correctly. The site team insists it was stored properly. And the client sees only delay and added cost.

There is also a balance to be struck between protection and expense. Products can be heavily wrapped, crated, or boxed to minimise risk, but this drives up transport costs and creates extra waste. Leave them exposed, and they face hazards like road debris, tar, grit, or weather damage, cheaper upfront but far more costly in remedials. The challenge is finding the right level of protection for each product type, route, and season.

This is why transportation cannot be treated as a simple delivery task. In MMC, where such a large proportion of the work is completed before the unit reaches site, transport is a quality-critical phase. It is not just a handover; it is a moving stage of production. Unless it is owned, inspected, and recorded in the same way as the factory and the site, quality will fail invisibly, and by the time the problem is noticed, it is too late.

Why Transport Is Not Just Logistics in MMC

In MMC, we are not moving bulk materials like sand, cement, or timber lengths. We are moving precision components, pods, panels, cassettes, service racks. Many are fully assembled, often fully finished. Some carry brittle finishes or pre-installed MEP; others rely on exact tolerances to connect seamlessly once on site. All are built with dependencies: pre-cut penetrations, fixed floor heights, programmed crane slots.

This makes the system fragile, not in the sense of weakness, but in the sense of sensitivity. A vibration across an unsecured lorry bed can loosen a bracket. A missed weather wrap can let moisture damage an otherwise sealed unit. A crane lift with the wrong rigging can stress or twist an internal frame.

The risk is not only obvious physical damage but the subtler impact of losing alignment. A few millimetres off tolerance may not be visible at dispatch but can cause major issues at installation. What looks fine on the trailer can quickly become a dispute when a module will not fit or a panel no longer lines up with its neighbour.

That is why transport in MMC cannot be treated as simple logistics. It is a quality-critical phase. Without the right checks, protection, and shared responsibility, the damage is only discovered later, argued over, and resolved slowly. With the right processes, planned securing, weather protection, inspection records, and clear chain of custody, transport can be a smooth extension of production quality rather than a weak link.

Build QA Ownership into the Transport Phase

Following the responsibility framework set out earlier, transport needs its own set of QA stages. These stages must have named responsibility, clear evidence requirements, and centralised documentation. Transport is not a neutral handover, it is a live stage of production, and quality ownership must be designed into it.

Transport as a Quality Boundary

Each of these steps is not just a task – it is a risk boundary. A missed strapping check is not minor, it can compromise an entire module. A missing arrival inspection leaves everyone exposed

Table 5.4 QA Responsibility Matrix for Transportation

QA checkpoint	*Responsible role*	*Evidence required*
Final pre-dispatch sign-off	Factory dispatch QA lead	QITC issued, photo evidence, QR scan
Load out inspection	Logistics coordinator	Strapping verification, bracing check, weather wrap
In-transit tracking	Haulier	GPS trace, no reroute flags
Arrival on site	Site logistics supervisor	Visual inspection, damage checklist, condition photos
Crane lift QA	Lift supervisor	Sling configuration photo, lift permit
Post, lift install readiness	Site QA lead	Interface verification, seal check, photo confirmation

to disputes. Unless the person responsible is trained, assigned, and equipped with the right tools, the whole process becomes fragile.

The shared QA tracker should be extended to cover transport. Each checkpoint must be logged with a photo timestamp, the name of the inspector, and any relevant notes. QR-tagged units make this even stronger by linking each physical component to its digital QA trail. When a unit is scanned at the factory gate, again at the site gate, and then after the lift, there is a single, joined-up chain of evidence.

If damage is found post-lift, the project no longer has a chain of excuses. It has a record showing where, when, and how the failure occurred. This does not only protect against disputes, it builds confidence, reduces delays, and reinforces the principle that in MMC, quality is owned end-to-end, not just inside the factory walls.

The Magnified Risks: Movement, Moisture, and Misalignment

Transport might create new risks, but it will certainly magnify the ones already present. What looks like a perfect product at dispatch can arrive compromised, not because of poor workmanship, but because of what happens on the road. The key is recognising where the risks lie and building checks that catch them before they become disputes.

- *Movement*: Even when units are strapped correctly, long pods or panels can flex during transit. That flex is often invisible until later, but it can crack tiles, shift pipework, or distort alignment points. Without a pre-lift and post-lift QA inspection, the damage may be blamed on the wrong party. With proper checks, the issue is flagged early and the chain of responsibility is clear.
- *Moisture*: Rain ingress through torn wrap, unsealed joints, or prolonged exposure can create hidden damage. The risks include mould, corrosion, or swelling of finishes. If a unit is left overnight on a lorry without an inspection protocol, there is no evidence of where the failure occurred. A simple moisture check or seal inspection on arrival can prevent a costly and contested dispute later.
- *Misalignment*: A panel or module that has warped slightly in transit may still be lifted and installed. But once that happens, every subsequent unit becomes harder to fit. Tolerances shrink, joints open up, and re-work escalates. At that point, the issue looks like poor installation, but the root cause was the journey. Embedding an interface check before installation protects against this spiral of error.
- *Incorrect bracing*: Many components rely on temporary bracing for safe transport. If that bracing is removed too early, or not checked before loading, the unit is at risk. Without a QA step between factory exit and site arrival, this decision goes unrecorded. When re-work becomes necessary, the responsibility is unclear, and the cost lands on the project rather than the party at fault.

Each of these examples reinforces the same point: transport is not neutral. It is a quality stage in its own right. By identifying the real risks, movement, moisture, misalignment, and bracing, and by setting clear inspection points around them, projects can transform transport from a source of hidden failure into a controlled and predictable part of the MMC workflow.

Site Readiness and QA Timing

One of the most common failures in MMC logistics is not the journey itself, but what happens when the unit arrives on site. A pod or panel might reach the gate in perfect condition, but if the crane is delayed or the slab is not ready, it sits exposed. Rain starts. The protective wrap is peeled back for inspection, then hastily rewrapped. Hours later, the lift finally happens, but by then no one is sure who checked what, or when.

This is why quality assurance in MMC is not just about what gets done; it is about when it gets done. Timing is as much a part of QA as the inspection itself. If the sequence is wrong, evidence is blurred, and disputes follow.

The logistics programme must therefore include:

- Allocated inspection time at each point. Do not squeeze checks into lifting slots; treat them as planned activities with proper duration.
- Named individuals with authority to hold or release the unit. Quality decisions cannot wait for a phone call up the chain; they need to be made in real time.
- Float for snag response, not just lifting windows. Build in time for corrective action if something is found. A programme with no slack for quality response is a programme waiting for delay.

These steps need to be embedded in the master programme, not treated as optional extras. They should appear on the Gantt chart as clearly as any install task, and they should be resourced and costed in the same way. When QA timing is visible, planned, and paid for, it stops being a bottleneck and becomes part of the delivery rhythm.

Handled this way, site readiness checks protect both time and quality. They ensure that every lift, every install, and every handover is done with evidence in place and confidence intact.

Risk Transfer: Who Owns the Damage?

One of the most persistent challenges in MMC logistics is the insurance gap. When a unit is damaged in transit, responsibility often sits in the grey space between contracts. The factory insists it left the gate in perfect condition. The haulier reports no incident on the road. The site team claims it arrived wet or misaligned. Meanwhile, the programme is slipping, and the argument distracts from fixing the problem.

The only way to resolve this is to define risk transfer points clearly in the contract and to tie them directly to QA documentation. This turns a potential dispute into a simple check of records.

Examples include:

- *Factory responsibility*: Covers condition until the unit is accepted by the haulier at the factory gate
- *Haulier responsibility*: Covers condition in transit until handover is signed by the site logistics lead
- *Site responsibility*: Begins once the lift is complete and documented by the site QA lead

These stages must be written into the contract, matched to inspection protocols, and signed off with evidence at every gate. A photo, a QR scan, a name, a timestamp, together these create a chain of custody that is recognised by insurers, understood by the project team, and defensible if a claim is made.

When risk transfer is visible, documented, and accepted, transport disputes stop being about opinion and start being about evidence.

Tools that Help (if Used Properly)

Technology makes it easier than ever to track MMC units across their journey. The tools are available. The challenge is embedding them consistently into the process.

- QR codes on each unit, linked to the common data environment (CDE), connect the physical product to its digital record
- Delivery photos taken on arrival, uploaded into a shared QA folder, create visual evidence of condition at the handover point
- Digital lifting permits, tied to unit ID and installer role, ensure each lift is authorised and traceable
- Weather logs and GPS trails record site exposure or unexpected delays, giving context for potential claims or remedials

These tools are not additional paperwork. They are replacements for the scattered spreadsheets, lost emails, and informal WhatsApp photos that often fail to stand up in dispute. But tools alone will not solve the problem. They must be backed by culture. That culture starts with naming transport QA roles at pre-start and following them through at every checkpoint.

Transport is not a gap. It is a live production stage. In MMC, quality is not only about how something is manufactured. It is about how it is handled. Transport is the stage where risk is most invisible, and where one small oversight can undo weeks of controlled factory work.

The answer is to build transport into the QA map as if it were a trade in its own right. Treat it like any other package: plan it, resource it, inspect it, and sign it off. Train site teams to inspect with intent. Equip logistics partners to document and defend. Define risk transfer in contracts. Use tools to build the golden thread through the journey.

When transport is treated as a live stage of production, not a forgotten gap between factory and site, the system holds together. And if something does go wrong, the evidence is there to resolve it quickly, not to fuel disputes that make it worse.

The TCA Process: Testing, Certification, and Accreditation

Modern Methods of Construction are often presented as faster, greener, and more consistent than traditional methods. The real test is proving those claims, to clients, funders, insurers, warranty providers, and regulators. That is the role of TCA: Testing, Certification, and Accreditation. These are not optional extras. They are how trust is built across a sector still learning how to standardise new ways of building.

TCA only works when it is integrated into the quality system, not treated as a prize at the end. Many current standards were written for conventional products and simple assemblies. MMC, by contrast, combines materials and constructs complete rooms, systems, or frames off-site. The task is therefore to demonstrate performance in a way that reflects how MMC is actually delivered, and to connect that proof to the quality assurance records outlined earlier in this chapter.

The aim is simple:

- Validate the promise with testing
- Prove the performance with QA
- Keep the golden thread through delivery and into operation

This section links directly back to responsibility and workflow and forward to Chapter 6 on time control. It also connects to the sustainability evidence in Chapter 4, where proof is required for any claims around carbon, circularity, or resource use.

What Are We Testing For?

In MMC, TCA applies at three distinct levels:

- Individual components, such as fire-resisting doors or acoustic panels
- Sub-assemblies, such as bathroom pods or MEP riser modules
- Full systems, such as volumetric units or wall cassette systems

Depending on the nature of the product, testing may be required for:

- Fire resistance and reaction to fire
- Structural strength and serviceability under load
- Acoustic separation and sound insulation
- Thermal performance, including U-values and airtightness
- Moisture resistance and water tightness
- Durability and performance under changes in temperature or humidity
- VOC (volatile organic compound) emissions where health and safety are relevant

In traditional projects, much of this testing is done on site or based on generic product data. In MMC, the timeline shifts upstream – tests are often carried out in laboratories or factories before the unit ever reaches site. This means the test data must be specific to what was actually built, not just to what was intended.

To be valid, results must be traceable to a batch, a bill of materials, and the conditions under which that batch was manufactured. A single pass test is not enough. There must be clear, auditable evidence that the products delivered to site were made under the same controlled conditions as the tested samples.

This is where the inspection plans become critical. A certified wall system still needs a factory QITP (Quality Inspection and Test Plan) that records key elements like fixings, sealants, adhesive types, densities, and cure times. A certified pod still needs documented proof that its penetrations, pipe runs, and membranes match the tested arrangement. Without that link, certification is just a label, it doesn't prove the product installed is the product tested.

The Certification Landscape

MMC products sit within a familiar but evolving landscape of marks, schemes, and certifications. The type of certification required depends on the product itself and the route it takes to market.

- UKCA marking is now the UK standard that replaces CE marking for most construction products. It confirms conformity to a relevant standard. For MMC products, this can be complex, as complete systems may span several categories or lack a single applicable standard.

- BOPAS (Buildoffsite Property Assurance Scheme) provides assurance on durability and process. It is often required by funders and mortgage providers for homes built using MMC.
- BBA certification remains a common route for recognition of both components and full systems, particularly for façades, walls, and roofing elements.
- NHBC Accepts offers a pathway for housing systems to gain warranty acceptance, which is critical for residential projects.
- ISO 9001 certifies quality management systems. While it does not guarantee product performance, it supports a culture of control, consistency, and continual improvement within the factory.

If a product claims sustainability performance, then certification must extend beyond safety and strength:

- BREEAM and LEED drive demand for environmental product evidence in both public and private sectors
- PAS 2080 is widely used in infrastructure for carbon management and can guide supplier-level evidence requirements
- Environmental Product Declarations (EPDs) are often required to prove the carbon impact of products per unit or per square metre
- Life Cycle Assessment (LCA) offers a full view of a product's environmental performance, from material selection through manufacturing, use, and end-of-life

These tools help project teams move from broad claims to measurable performance. They also support procurement decisions. As shown in Chapters 3 and 4, without verified data, it becomes harder to specify, gain approval, or secure funding.

In practical terms, a product with a current EPD and a clear LCA pathway has a stronger chance of being accepted on frameworks with net zero or circularity targets. The proof doesn't just support quality, it supports delivery.

Certification Is Not Control

Many projects mistakenly assume that holding a certificate guarantees quality. It does not. Certification only confirms that a system can meet a standard, when it is built to the approved design, using the right materials, and following the correct process. It does not guarantee that this unit, on this lorry, for this building, meets that same standard today.

Real control lives in the quality system. A certified wall or façade system still needs a QITP and QITC that confirm the factory followed the same build-up, materials, and fixings as the tested sample. A bathroom pod with a durability certificate must be backed by batch records showing inspections completed, defects identified, and corrections made. If a product carries an EPD, the exact material grade and mix must appear in the bill of materials and the delivery notes, so that any later substitution does not quietly undermine the claim.

This is the line between assurance and theatre. Certificates are useful signals, they open doors, support approvals, and build initial trust. But they do not replace control. Quality comes from clear roles, workflows, and templates that build the chain of evidence from design through to handover. This section shows how certification must be plugged into that chain, so every performance claim is backed by proof, and every installed unit can be traced to the system that earned the certificate.

An Integrated TCA Workflow for MMC Projects

To embed certification into delivery, not just into bids, the project can follow a simple sequence. It mirrors the design freeze logic in Chapter 6 and the QA workflow from earlier in this chapter.

Every step builds accountability. Every document builds traceability. In Chapter 6 this pays back in time. Clear evidence reduces disputes and speeds decisions. In Chapter 7 on cost, this reduces risk allowances and claim exposure.

TCA is How MMC Proves It Works

Testing, certification, and accreditation (TCA) build confidence that the asset meets the expected standards for safety, performance, and sustainability. They reduce the likelihood of late design changes and unplanned re-work. They shorten decision time when queries arise. They also support later-stage processes like valuation, insurance, and regulatory sign-off. In housing, for example, TCA helps mortgage lenders and warranty providers accept MMC systems. In sectors like health and education, it strengthens approval routes that now expect robust, product-specific evidence.

For clients and funders, the benefit is not only technical, it is commercial. Certainty of performance reduces contingency allowances. Certainty of process reduces delays. As shown in Chapter 6, this leads to a more stable programme. In Chapter 7, the same certainty supports clearer pricing structures and fewer claims.

Table 5.5 TCA Workflow for MMC Projects

Stage	*Purpose*	*Key actions*	*Evidence or output*	*Primary owner*	*RIBA stage*	*QA or TCA link*
Product selection	Fix the exact system	Confirm variant and limits of use	System ID, datasheet extract	Design lead	2–3	Basis for TCA pack
Certification review	Check approvals match design	Collate certs and test reports	TCA pack index	Technical manager	3–4	Scope of approval set
QA input	Embed acceptance criteria	Add TCA refs to QITP	Updated QITP template	QA lead	3–4	Hold points defined
Production QA	Prove like for like build	Record batches, dimensions, tests	Batch log, photo set	Factory QA	4–5	Evidence of conformity
Post production	Confirm unit meets approval	Issue QITC for each unit or batch	QITC with unit ID	Factory manager	5	Ready for dispatch gate
Handover pack	Make records accessible	Compile certs and QA logs in CDE	Handover folder link	Principal contractor	5–6	Golden thread bundle
Asset register	Enable future verification	Tag assets with IDs or QR codes	Asset list with TCA refs	FM lead or client	6–7	In use traceability

TCA is often seen as a compliance task. In MMC, it is part of delivery. It allows the project team to prove what was promised. It provides a shared language across design, factory, transport, and site. And when joined to tools like QITPs and QITCs, it becomes a living record, not just a static file. Used well, TCA builds trust, accelerates decision-making, and helps turn repeatable products into repeatable outcomes.

Product Quality as a Project Culture

This chapter closes with the most important point – quality is a culture, not a checklist.

In traditional projects, handover often marks a clean break. The contractor leaves, the client takes over, and a new team lives with the result. In modern methods, that line is blurred. The product is designed as a system, manufactured as a system, and often supported as part of a long-term asset plan. What happens after handover is not a separate phase, it is the final test of everything that came before. If the quality system was real, the building performs. If it was just paperwork, small issues turn into daily frustrations.

The shift we are asking for is simple to say and hard to fake – quality must become everyday behaviour.

That means every person knows what a pass looks like, what is checked at each stage, and what is not acceptable. It also means people understand why the checks matter. When a factory operative knows how a missed seal leads to a leak weeks later, they speak up. When a site installer sees how a minor misalignment breaks the fire or acoustic line, they stop the lift and call it out. Early reporting is not a weakness, it is strength.

To make this real, start with a proper induction. Bring the factory teams, site crews, logistics leads, and designers into the same room. Walk through the full process: from design freeze, to factory gate, to transport, to crane lift, to commissioning. Show the product in use. If it's a pod, explain how it connects, drains, breathes, and is maintained. If it's a cassette, show how the joints and seals protect fire stops and acoustics. This is not a toolbox talk. It is onboarding for a system.

Make quality visible every day. Use a simple dashboard in the factory and on site: first-time pass rate, number of open defects, time to close defects, and the most common causes. Discuss them in daily huddles. Put a near-miss board where everyone sees it, record interface misfits, packaging errors, sequencing clashes. Treat them like learning points, just like health and safety. And when someone raises an issue, close the loop. Show what was done. Share the learning. People keep reporting when they see it makes a difference.

Leadership sets the tone. If quality is treated like a tick-box, so will everyone else. But if leaders attend inductions, ask for evidence in meetings, stop poor installs, and defend time spent to do it right, the culture follows. Build this into performance incentives. Reward proactive snag reporting and first-time pass, not just hitting dates. Include quality in subcontractor reviews. Make it clear – a clean handover is as valuable as a fast one.

Join the systems together. Testing, certification, and accreditation show what a system can do. The inspection plans prove what this unit did. Transport checks protect what left the factory. Site checks confirm that what arrived matches what was designed. The golden thread is the set of records that links these together. Use the common data environment. QR-tag each unit. Store photos and sign-offs against the unit ID. Avoid evidence stuck on phones and email threads. When records sit in one place, decisions are faster, disputes are fewer, and trust builds.

Plan quality in time. This is the bridge to Chapter 6. A fast method still needs control. Inspection time must be built into the programme, just like lifts and pours. Name the people who can hold or release a unit. Add buffer time for snag response. When QA time is visible on the programme and valued in payment, it doesn't get squeezed. The result: fewer surprises and smoother delivery.

Protect accountability. Name the people responsible at each gate: factory dispatch, load-out, arrival, lift, install, commissioning. Give them the tools to make and record decisions. Use one shared tracker. If damage appears, rely on a chain of custody, not a chain of excuses. This protects both the programme and honest suppliers.

Design for maintenance. Quality does not stop at practical completion. Walk the routes your facilities team will use. Can a filter be changed without cutting finishes? Can valves be reached without removing a panel? Can a pod be isolated and repaired without closing a whole floor? These decisions often cost little if made early, and save money for years. This is where the golden thread proves its worth.

Keep learning. After handover, track the data – defects per unit, mean time to fix, the top three recurring issues. Feed that back into the next design freeze and the next factory plan. As Chapter 4 argued, supply chain maturity grows through feedback loops. The goal is not zero defects on day one. The goal is fewer and smaller defects on every project, because the system keeps improving.

Keeping the tone positive. This book is honest, not critical. Modern methods can deliver excellent quality, but only if the system is designed and run well. We've shown how to do that. Use the inspection plans. Use the responsibility matrix. Use the transport checks. Use the certification workflow. Use the programme gates. These are practical steps, not theory.

This chapter has focused on product quality. Chapter 6 carries the same logic into time. It shows that real speed comes from real planning. Quality and time move together. Good control gives predictable programmes. Predictable programmes reduce cost risk (see Chapter 7).

Finally, remember that the real test of modern methods is not handover day. It is year one, year five, and year 15. If we get quality right, operations are calmer, customers are happier, and money isn't wasted on repeat work. If we don't, we simply move old problems into a new method. *The choice is ours.*

Chapter 6

Product Time Management

Introduction: The Paradox of Speed in MMC

Chapter 5 ended with a clear message: factory production offers strong potential for quality, but only when supported by the right systems, processes, and accountability. The same applies to time. In Modern Methods of Construction (MMC), time is not simply a number on a programme chart, it is a system that must be shaped around the product.

MMC is often promoted as faster. Quicker to build. Quicker to finish. Quicker to hand over. And in many ways, that's true. A bathroom pod made in a factory can be delivered and installed in a matter of hours. A structural panel arrives with insulation, finishes, and services already in place. Modular buildings can rise rapidly, with superstructures going up in days, not months. These examples are real, and they demonstrate the potential for impressive gains in site productivity.

But there is a hidden side to that speed. What's often left out of the conversation is that MMC compresses time on site by demanding more time before site work begins. This is the paradox: while you save time during the build, you spend more time preparing for it. If that preparation isn't done properly, the benefits of speed can vanish, or worse, become risks.

In traditional construction, many decisions can be made later. Wall layouts can be adjusted after groundworks start. Kitchens can be re-specified mid-build. Procurement can be phased. That flexibility absorbs indecision. But MMC doesn't allow for that. The modules or panels are manufactured early. That means they must be fully designed, signed off, and sequenced before they arrive on site. Change after manufacture is not just inconvenient, it can be impossible or extremely expensive.

This is the reality of time in MMC: speed is front-loaded. Good time management means starting earlier, locking down key decisions sooner, and planning every step in greater detail. It's not just about building quickly. It's about ensuring the whole system is ready to move without interruption. If approvals, logistics, or design freeze slip, the consequences are immediate. A completed unit may be delivered on time but sit idle because the crane is late, the slab is out of tolerance, or the access route is blocked.

This chapter explores how to manage that complexity. It argues that time in MMC is not a by-product, it's a discipline. Managing time means managing coordination, certainty, contracts, tools, people, and risk.

We begin by comparing how traditional and MMC programmes are structured. In traditional builds, tasks follow a largely linear path: design, then procurement, then construction. There are buffers between phases, and room to recover. MMC operates differently. It requires parallel workflows, where manufacturing begins off-site while foundations and infrastructure are still

DOI: 10.1201/9781003570431-6

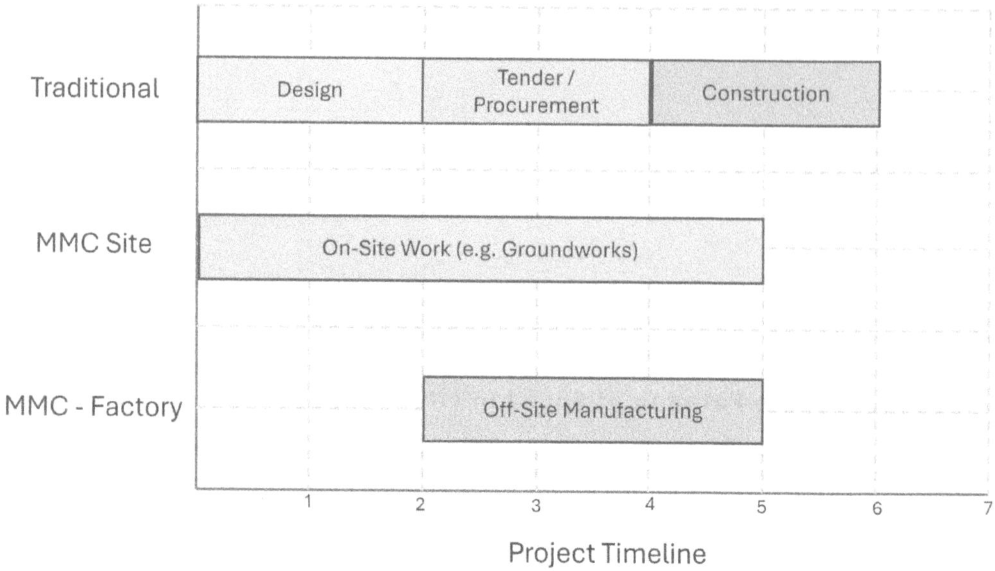

Figure 6.1 Production Timeline Comparison

in progress on-site. Done well, this overlap reduces the overall programme. But done poorly, it introduces friction. A delay in design, factory approvals, or groundworks can halt progress across the whole stream. Coordinating these moving parts requires careful logic, real-time communication, and a clear understanding of dependencies.

From there, we turn to design freeze, one of the most critical but often misunderstood milestones in MMC. This is the point where the design must be finalised to allow manufacturing to start. If the freeze is missed or not respected, the risk of costly re-work increases dramatically. We examine what a good design freeze looks like, who owns it, and how to protect it in project governance.

Next, we address contracts. MMC changes how time risk is managed commercially. Factory slots must be booked early. Delivery dates must align with crane availability and installation teams. Payment schedules may need to track off-site progress rather than traditional site milestones. If the contract does not reflect this reality, time pressure quickly becomes cost pressure. This section explains how contracts must adapt to MMC's different time logic.

We then explore the tools that support time control. Digital programmes, 4D BIM, production calendars, logistics dashboards, and gate checks can all add value, but only if they are actively used and properly integrated. A 4D model that is never updated does not save time. A delivery tracker that ignores site readiness causes delay. This section focuses on how these tools should be embedded into live management, not just used at tender or reporting stage.

Another critical piece is people. Even the best systems fail if teams don't understand how to use them. Factory managers need to understand the sequence of install. Site managers need to understand how factory changes ripple downstream. Designers need to understand what decisions must be locked, and by when. This section highlights the training, communication, and leadership needed to make time systems work in practice, and shows how assumptions and poor handovers often lead to lost time.

Finally, we look at the link between time, cost, and quality. Faster is not always cheaper. Rushing can cause errors that are expensive to fix. But predictable timeframes do improve cost certainty. For public clients like NHS trusts or education bodies, that reliability can be just as valuable as speed. This section connects time logic to the cost themes in the next chapter, including how MMC changes value engineering, budget setting, and supply chain coordination.

The chapter closes by reframing time as more than a benefit. In MMC, time is a responsibility. If we want to build faster, we must plan better. If we want site time to be smooth, we must invest more in preconstruction, procurement, and sequencing. Speed is not about cutting corners, it's about being ready to move when the moment comes.

We do not build faster by rushing. We build faster by being ready.

Time in Traditional vs MMC Construction

The most visible promise of MMC is speed. But to understand where time is gained, and where it is put at risk, we must first understand how time behaves differently in traditional and MMC-based projects. This section compares the two models: how work is sequenced, where decisions sit, and how time is either protected or exposed. It shows that the real time savings in MMC come not from rushing, but from removing overlap, waste, and repetition.

Labour-Intensive, Weather-Sensitive, Trade-Dependent

Time in traditional construction is shaped by its process. It follows a mostly linear path: design first, then procurement, then build. On-site, the work progresses in sequence, foundations, then structure, then interiors. Each trade depends on the previous one being finished. This seems logical and manageable in theory. But in practice, it introduces layers of fragility. Every step relies on the one before being done correctly, on time, and in the right conditions.

Traditional construction is labour-intensive and weather-sensitive. Most of the work happens outdoors, exposed to rain, wind, cold, or heat. Wet trades like screeding or plastering are directly affected by temperature and humidity. Delays are common. Materials may arrive late. Access routes may be blocked. Specialist trades might not be available when needed. These risks are often hidden in contingency plans, extra time called "float", that sit quietly in the programme. But they represent real uncertainty.

Skilled labour availability is another constraint. Traditional builds rely heavily on trades like bricklaying, joinery, plumbing, and roofing. If a key team member is unavailable, progress slows. Replacing site-based skills isn't easy, especially for complex or bespoke work. Some tasks are reliant on one person's experience or know-how. Quality often depends on who is on the tools that day.

There is also a coordination problem. Because many trades work in the same spaces, clashes occur. A dryliner may be ready to work, but the electrician isn't finished. A kitchen fitter arrives to find damp walls. Even with detailed programming, the on-site reality is fluid. Teams adjust and work around each other, but this introduces inefficiency. Each workaround adds risk, both to time and to quality.

Even repeat tasks vary. A plasterer may do the same job differently in every room. A door may be hung slightly out in one unit and perfect in the next. There is little standardisation, which makes it harder to track performance or improve productivity over time.

In traditional builds, time is reactive. Programmes flex around issues, but often at a cost. Rework is common. Quality varies. Progress is measured by how many trades are on-site, not by how much usable work is completed. Delays are absorbed by effort, longer hours, more labour, more supervision, but not always with better outcomes.

This fragmented approach explains why traditional construction has struggled to improve productivity. While other industries have gained efficiency through automation, standardisation, and lean processes, much of construction still depends on people building in changing conditions, by hand, one task at a time.

MMC Programme Characteristics: Running Factory and Site in Parallel

Modern Methods of Construction offer a different approach. Instead of building everything on-site, MMC moves key activities into a factory. Modules, panels, pods, or other pre-manufactured components are made in controlled conditions, often while the site is still being prepared. This allows two major workstreams, factory and site, to run in parallel, not in sequence.

This parallel approach is the core reason MMC can shorten programme durations. A bathroom pod, which might take two or three weeks to fit out on-site across multiple trades, can be completed in the factory in a few days and installed in a few hours. Wall panels can arrive pre-insulated, pre-finished, and ready to install. Entire volumetric modules can be delivered fully complete, reducing the need for internal work on-site. This creates huge potential to reduce programme time, but only if the system around it is properly managed.

The key difference is that MMC time savings come from coordination, not flexibility. Unlike traditional builds, there is little room to shuffle tasks or adjust on the fly. Once the pod is built, it cannot be modified easily. Once the panel is made, the window size is locked. If the slab is not ready when the delivery arrives, there may be no safe place to store the unit. If the crane is not booked, the module cannot be lifted. The sequencing is tight, and the dependencies are fixed.

This creates a different type of time risk. In traditional builds, time is lost during construction. In MMC, time risk moves upstream, into the design phase, the factory programme, and the logistics schedule. The critical path often begins earlier than teams expect. Mistakes that would be minor on a traditional site can have major implications in MMC. A mismatch in MEP connections may not be noticed until modules are on site, at which point, fixes are difficult and expensive.

That's why design freeze matters so much. The manufacturing programme can only begin once the design is finalised. And once manufacturing starts, changes become high-risk. Decisions that might have been left open in a traditional build, wall types, bathroom layouts, electrical positions, must be locked earlier. This can feel uncomfortable to teams used to late-stage decisions or continuous design development. But it's essential for programme certainty in MMC.

This also means that the success of the programme is no longer driven by the construction team alone. Factory teams, logistics managers, crane coordinators, and digital schedulers all become critical to keeping time on track. A missed sign-off can delay production. A failed inspection can hold up dispatch. Poor weather at site can delay a lift and back up deliveries.

The whole system becomes tighter. Faster, yes, but less forgiving. The answer is not to rush, but to plan with greater accuracy, commitment, and communication. This includes:

- Coordinating digital models between design and manufacture
- Locking interface points between factory and site
- Sharing production calendars and delivery schedules with the whole team
- Tracking QA sign-offs and approvals in real time
- Integrating factory milestones into the main programme, not as external events

Used well, this approach removes waste, repetition, and delay. It brings forward decisions but locks in outcomes earlier. Done poorly, it can amplify problems, because errors are no longer caught and corrected on-site, but arrive fully built and hard to fix.

Summary: Two Models, Two Mindsets

Traditional construction is based on sequential progress and reactive flexibility. MMC is based on parallel streams and proactive coordination. Both models require skill and planning, but they behave differently when it comes to time.

In traditional builds, time can appear flexible, but that flexibility often hides inefficiency. In MMC, time appears rigid, but that rigidity enables speed, if managed well.

To benefit from MMC's speed, project teams must change how they think about time. Early decisions matter more. Dependencies must be visible. The site and the factory must operate as one team. And time must be protected not just during construction, but from the first moment of design.

This shift is not cosmetic. It is fundamental. Because in MMC, time saved on site is earned in design. And if that time is not invested early, it will cost more later.

Simplified Sequencing, Fewer Surprises

One of the most powerful advantages of MMC is the simplification of the construction sequence. By shifting complex, multi-trade tasks off-site and into the factory, MMC eliminates many of the small delays and coordination issues that traditionally slow projects down. Drying times, trade overlaps, and material handling bottlenecks are reduced. Instead of five trades completing a bathroom over two weeks, a factory can produce a fully fitted bathroom pod in a matter of days, ready for single-day installation on site.

This repeatability brings structure and certainty. Each module or cassette follows the same design, the same process, and the same inspection routine. Fewer crane lifts are needed. Scaffolding may be reduced or avoided altogether. Site crews follow a predictable routine, position, connect, commission, rather than managing dozens of small, variable tasks. The result is smoother progress, fewer site clashes, and less reactive decision-making.

MMC also brings safety and logistics benefits. With fewer trades in the same space, the site is less congested. There is less cutting, mixing, and working at height. This reduces both risk and delay. Fewer deliveries and less on-site waste mean cleaner, safer working conditions. When properly planned, this creates a virtuous cycle: better safety supports better productivity, which supports better time performance.

But none of this happens by accident.

To unlock the benefits of simplified sequencing, planning must be exact. Factory-made elements must match site conditions precisely. Foundations must be built to tolerances. Openings must align. Lifting and access routes must be confirmed in advance. Quality checks must be passed before dispatch. Transport slots must be secured, and the site must be ready to receive. If even one part of this chain is missed, the knock-on effect can be significant.

This is the central lesson: MMC reduces many of the traditional risks, but only by introducing new dependencies that must be managed earlier and more rigorously. Time is saved not by rushing, but by coordinating. Fewer surprises happen on site, but only if every step before site is executed with discipline and precision.

Why Faster Still Needs Buffers

It's easy to say MMC is faster, and, in many ways, it is. But the speed is not evenly spread. It is not that every task is shorter, but that the entire system is more controlled. Tasks happen in parallel. Work is completed off-site while foundations are being built. Installation is faster because the product arrives pre-finished. The schedule is tighter, but it depends on more things going right at once.

This creates a new challenge: what happens when something goes wrong?

In traditional construction, if a material delivery is late or a task is delayed, the team can often reshuffle tasks or bring in extra labour. There's some slack in the system. With MMC, the margin for error is smaller. A missed production slot can delay dispatch by a week. A failed quality inspection may stop a module from leaving the factory. A 10mm site error may prevent a pod from fitting at all. The result is a cascade of knock-on delays.

That's why fast programmes still need buffers. In MMC, these buffers are not signs of weak planning, they are signs of smart planning. They absorb small shocks before they become major disruptions.

Examples of MMC buffer zones include:

- *Pre-production buffers*: Additional time to confirm all specifications and interface details before factory orders are locked.
- *Module-to-module coordination time*: Time for dry runs, interface mock-ups, or design coordination between adjoining factory units.
- *QA and sign-off time*: Time built into the factory schedule for quality checks and documentation.
- *On-site adjustment windows*: Short periods to address minor misalignments or non-conformities during installation.

Another critical area is storage and logistics. The ideal MMC model is Just-in-Time (JIT): components arrive on-site only when needed, reducing storage costs and damage risk. But in the UK, confidence in JIT delivery is still developing. Some contractors ask for early delivery and local storage as a precaution. This introduces new costs, handling risks, and site congestion. It also undermines the efficiency MMC is designed to create.

The root cause is often weak coordination between factory, logistics, and site. Better digital tracking, clearer milestones, and shared production calendars are needed to build trust. Until then, buffers remain a necessary safety net.

Ultimately, the point is this – MMC speeds up construction by compressing time, but compression creates fragility. A fast programme without buffers is not efficient. It is brittle. One missed task can unravel the entire sequence. A smart MMC schedule includes enough slack to handle real-world variation, without losing momentum.

Visual Case Study: Tide Construction's Ten Degrees, Croydon

- Tide Construction's Ten Degrees development in Croydon offers a clear example of how MMC can dramatically change time performance.
- The scheme comprises two towers, standing at 38 and 44 storeys, reaching 135 metres in height. Delivered in just 26 months, the project includes 546 build-to-rent homes, approximately 50% faster than the timeline typically associated with high-rise traditional builds.

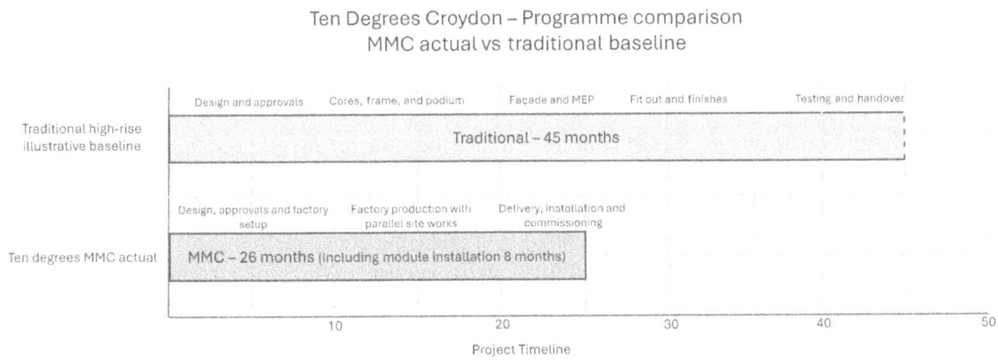

Figure 6.2 Case Study Comparison

- Key to this speed was the use of 1,526 volumetric modules, built off-site by Vision Modular Systems. Each unit was pre-fitted, quality-checked up to five times, and delivered to site fully finished, including kitchens, bathrooms, and internal finishes.
- The factory production ran in parallel with groundworks. As soon as the podium levels were complete, modules were craned into place on a tightly coordinated schedule. The final module was installed just 35 weeks after the first lift, ahead of programme.

What made this possible?

- Factory precision allowed repeatable units to be stacked rapidly with minimal on-site work
- QA processes ensured that installation would not be delayed by defects
- Digital tracking meant every module's location, inspection status, and delivery time were known in real time
- The team committed to design freeze and held the line, reducing late-stage design risk

This was not just a fast project. It was a controlled project. The speed came from preparation, coordination, and trust in the process. The lesson from Ten Degrees is not that MMC can go faster. That's already known. The lesson is that it only goes faster when every part of the system is aligned, factory, site, logistics, design, and client.

Design Freeze and the Front-Loading of Time

What is Design Freeze, and Why Does it Matter More in MMC?

In traditional construction, design often continues to evolve well after construction begins. Minor changes, late client decisions, or incomplete coordination can be absorbed by the trades on site. While not ideal, this approach is manageable because of the open-ended, reactive nature of site work. Adjustments can be made without halting the entire programme.

Modern Methods of Construction (MMC), however, do not offer the same flexibility. Once a wall panel is manufactured, or a bathroom pod is assembled, that product is fixed. Changing it after the fact is rarely safe, cost-effective, or practical. A small modification may trigger a full rollback – revisiting the design, remanufacturing the unit, repeating quality checks, rescheduling delivery, and potentially disrupting the installation sequence.

This is why design freeze is a critical milestone in MMC. It's not just a drawing deadline, it's a programme control point. If the design is not frozen, the factory cannot build. If the factory cannot build, the module cannot ship. If the module cannot ship, the site cannot install. The entire chain halts.

A timely and clear design freeze enables off-site manufacturing and on-site construction to proceed in parallel with confidence. It also carries legal and commercial implications – once a manufacturing order is placed, the specification becomes contractual. Any change after that point is treated as a variation, with associated cost and time consequences.

Design freeze also reinforces quality. It locks in the version of the product that the factory will inspect, that will be logged in the digital record, and that will be verified on site. It strengthens the golden thread outlined in Chapter 5 by creating a stable, certifiable version of the building components before production begins.

For the freeze to work, ownership must be clear. Without defined responsibilities, the freeze can slip, and that slippage cascades into delay. At a minimum, the following roles must be clear and active at the point of freeze:

- Client – confirms scope, layouts, and finishes
- Design Lead – confirms coordination, tolerances, and compliance
- MMC Manufacturer – confirms production inputs, material choices, and testing requirements
- Principal Contractor – confirms site readiness and programme alignment
- Quality Lead – ensures inspection plans and certification (including TCA) are aligned with the frozen design

When these roles are aligned, the design freeze becomes a stable and enabling milestone. When they are unclear, decisions drift and factories wait. The result is lost time that cannot be recovered later. A strong design freeze does not delay progress, it makes MMC speed possible.

Late Design Changes Create Exponential Delay in MMC

As noted in Chapter 5, late changes in MMC often lead to disproportionate increases in cost. The same principle applies to time. A seemingly small design update made late in the process can lead to a much larger delay due to the tightly connected nature of design, manufacturing, logistics, and installation.

In traditional construction, a late design change, such as moving a sink, might delay plumbing or tiling for a few days. Site teams can usually reshuffle tasks, bring in extra trades, or continue work elsewhere.

In MMC, there is far less room for manoeuvre. If the change affects a component already in production, the entire factory line may need to pause while new drawings are prepared. If physical work has started, existing units may need to be stripped out, modified, or even scrapped. Transport bookings are missed, site teams are stood down, and crane slots are lost.

What happens when a change is made after production has started?

- The factory halts the batch for updated drawings and instructions
- Part-built units are modified, re-worked, or discarded
- New parts must be sourced and fed back into the production line

- Quality checks are restarted for the revised units
- Transport schedules are missed and must be rebooked
- Crane slots and install crews are delayed or rescheduled
- Follow-on trades are pushed back, sometimes by weeks

Each step may only add a day or two, but together, they multiply. A simple design update can cause a ripple effect through factory, logistics, and site that adds weeks of delay.

A simple example:

Twenty bathroom pods are in production. A late request changes the basin location. The production line halts for two days to revise drawings. Ten pods already built require re-work, adding another five days. Delivery is missed, requiring a new transport slot a week later. The crane booking is also lost, pushing installation back another four days. Install crews are rebooked, and follow-on trades lose a further week.

Total delay: Three weeks or more, triggered by a seemingly minor change.

Why delays grow so quickly in MMC:

- Factories work in batches, following fixed production schedules
- Quality gates prevent release without complete documentation
- Transport slots, road permits, and crane bookings are limited and inflexible
- Multiple trades depend on accurate installation windows
- Re-work must be verified before progression
- If late change is unavoidable:
- Use a formal change control process with clear approval routes
- Carry out rapid impact assessments across design, factory, logistics, and site
- Prioritise only essential changes for compliance or safety
- Place non-critical changes into future batches or phases
- Update all relevant parties within one working day

Late changes in MMC are not just design updates. They trigger a full reset across several parts of the programme. The time cost is rarely measured in days, it is often measured in weeks.

The best defence remains a firm design freeze backed by defined responsibilities and agreed evidence. In MMC, speed depends on decision stability. Once production starts, change becomes delay.

Design Freeze in the RIBA Plan of Work with the DfMA Overlay

The Design for Manufacture and Assembly (DfMA) Overlay to the RIBA Plan of Work 2020 offers a practical framework for managing design freeze in MMC projects. Its key message is simple: MMC only delivers time benefits when decisions are made early, and control points are clear.

When should the freeze happen? In most MMC projects, design freeze should be complete by the end of RIBA Stage 4 (Technical Design). For complex or highly integrated components, certain freezes, particularly for Tier 1 and Tier 2 elements, may begin during Stage 3 (Spatial Coordination). Early supplier engagement supports this, allowing teams to confirm tolerances, interfaces, and test plans ahead of time.

A Tiered Approach to Design Freeze

The DfMA Overlay promotes a component strategy, a structured way to freeze different layers of the design at the right time. A simple three-tier system is effective:

- Tier 1 – Structural geometry, layout, primary interfaces
- Tier 2 – Internal coordination, MEP routes, service penetrations
- Tier 3 – Finishes, fixtures, internal fit-out

Each tier affects different areas of the programme:

- Freezing Tier 1 allows structural and groundworks teams to align with factory needs
- Freezing Tier 2 enables accurate placement of openings, ducts, and fixings
- Freezing Tier 3 allows materials and finishes to be ordered without risk

How this maps to RIBA Stages:

- Stage 2 – Define MMC ambition and identify manufactured components
- Stage 3 – Complete spatial coordination to match factory tolerances
- Stage 4 – Freeze Tier 1 and Tier 2 for off-site packages; confirm production readiness
- Stage 5 – Monitor delivery and hold to the frozen design; avoid significant change

If the design is not frozen by the end of Stage 4, the project enters manufacturing with uncertainty. This creates time risk that cannot be recovered on site.

NHS Toolkit Example: Full Business Case Readiness

The NHS MMC Assessment Tool provides a useful model of how to assess freeze readiness. At the Full Business Case (FBC) stage, the toolkit asks:

- Has design maturity been achieved for MMC components?
- Are coordinated drawings and specifications in place?
- Have suppliers been engaged and confirmed?
- Is the Pre-Manufactured Value (PMV) aligned to the procurement plan?

In simple terms, this is a design freeze audit. It asks whether the project is ready to manufacture with confidence.

The NHS approach reinforces three essential time enablers:

- *Clear component strategy*: Defines scope and sequencing
- *Pre-certification and QA alignment*: Reduces re-work and accelerates sign-off
- *Logistics planning*: Matches factory output to site readiness

These principles connect directly to Chapter 3 on procurement, Chapter 4 on supply chain readiness, and Chapter 5 on quality and the golden thread. They also lay the foundation for Chapter 7 on cost, where predictable design and programme stability help reduce claims and manage risk.

In summary, the DfMA Overlay transforms design freeze from a single milestone into a structured set of decisions. When teams follow this framework, the factory can plan with certainty, site works can proceed smoothly, and the full-time advantage of MMC can be realised.

Managing Design Freeze

Managing the design freeze in MMC is not about chasing a date on a programme. It is about making sure that decisions are ready, owned, and understood at the right time. The aim is not to block change for its own sake. The aim is to lock down what the factory needs, when it needs it, so that both factory and site can move with confidence.

A design freeze is not just a technical milestone. It is a delivery tool. It allows parallel progress, protects quality, and limits re-work. To manage it well, teams need early clarity, visible decisions, and shared discipline.

Start with the Client and Users

Many late changes come from uncertainty, not indecision, but misunderstanding. Engage the client and users early. Hold decision-making workshops. Walk them through layouts using simple visuals, mock-ups, or virtual walkthroughs. Focus on practical choices: room layouts, fixture positions, finishes.

Be clear: once the freeze is set, changes will carry both time and cost impact. Link this to Chapter 7 so that the client understands the commercial effect of design movement after the freeze.

Use a Simple Decision Map

Build one page per MMC package showing the decisions that drive design, production, and procurement. For example:

- Bathroom pods: layouts, sanitaryware, MEP connections
- Façade panels: window types, finishes, fire stopping
- Risers: service routes, access, fire compartmentation

Mark which decisions are critical and when they must be locked. Place this on the master programme and include it in coordination meetings. This supports both programme logic (this chapter) and quality assurance (Chapter 5).

Create a Freeze Control Log

Keep a single, live record of freeze status. For each MMC package, the log should include:

- What is frozen
- The date it was frozen
- Who approved it
- Drawings and models included (with revision numbers)
- Any agreed tolerances or flexibilities
- Any variations raised after the freeze

Store this log in the Common Data Environment (CDE) so that the factory, site, and client are always working from the same source.

Communicate What Is Fixed

When a freeze occurs, issue a short freeze notice. Don't rely on meeting notes or assumptions. Use a simple template that includes:

- Drawing and model numbers with revisions
- Confirmation of effect on procurement and programme
- Statement of any known risks or tolerances

This ensures consistency and reduces downstream disputes. It also enables site and factory teams to act with certainty.

Check Interfaces Before You Freeze

Most programme delays stem from missed interfaces, particularly between MMC components and traditional elements. Before freezing a package, run three essential checks:

- Digital clash detection for structure, pods, risers, and openings
- Tolerance review against manufacturer guidelines
- Mock-up or "first off" unit, if possible

Capture this evidence in the inspection and test plan. This links back to the golden thread (Chapter 5) and the DfMA overlay. A successful freeze is one that holds under pressure, and that only happens when interfaces are tested before the lock-in.

Set a Post-Freeze Buffer

Hold a short, controlled buffer after the freeze for tidying up borderline issues. Typically, two to three weeks is enough. This should sit before factory release, not at the end of the programme. Make the buffer visible, on the programme, in meetings, and in dashboards, so that it is actively managed, not silently consumed.

Run Clear Governance

Hold a weekly freeze readiness meeting for active MMC packages. Use a one-page tracker for each package showing:

- Design maturity status
- Interface check completion
- Approval status
- QA documentation readiness
- Go/No-go decision for factory release

Name people, not companies. Include direct contacts. Ambiguity delays decisions. Clear ownership drives action.

Define an Exceptions Process

Some late changes are unavoidable. Define simple rules to allow justified exceptions:

- Safety or statutory compliance
- Planning or landlord conditions
- Errors that prevent use or certification

All other changes should be deferred to the next batch or to a later phase. Use a short change form to assess impact across time, cost, factory, logistics, and site. Make the decision within one working day to prevent drift.

Align with Contracts and Procurement

Freeze dates should appear in key contracts and supplier orders.
They should link to:

- Factory production slots
- Delivery windows
- Pre-dispatch inspections
- Crane bookings and access windows

When freeze points are contractual, risk allowances reduce, and suppliers can price with more confidence (see Chapter 3).

Make It Visible

Use visual management in both the site office and factory office. A simple wall chart with packages marked green, amber, or red for freeze readiness is enough. Use a shared calendar showing key milestones:

- Design release
- First-off production
- Inspection gate
- Delivery and install window

Visibility prevents surprise. It also allows early warning when a freeze is slipping.

Link Freeze to Quality and Programme Control

Once frozen, the approved design should flow into the inspection and test plan. Every component should be traceable to the frozen version using:

- QR codes
- Digital checklists
- Photographic records

This keeps the golden thread intact and ensures disputes at handover are reduced. It also keeps the programme honest, and teams install exactly what was signed off.

A Simple Flow That Works

- Agree MMC scope and engage suppliers during RIBA Stages 2 and 3
- Map critical decisions and interfaces by package
- Complete Tier 1 and Tier 2 information by end of Stage 4
- Run interface checks and complete a mock-up or first-off unit
- Issue freeze notice and update the control log
- Hold a short buffer for tidy up post-freeze
- Release to factory with QA documentation in place
- Monitor delivery and installation against the frozen design in Stage 5

Why This Matters

A firm design freeze does not slow a project. It is the foundation that allows factory and site to operate in parallel, without conflict, error, or drift. It reduces late change, protects time, and supports quality assurance. In MMC, design freeze is the line between control and disorder.

When we set the freeze early, manage it clearly, and protect it fairly, the programme gains the speed that MMC promises, and the project team gains the confidence to deliver.

Contractual and Commercial Implications of Time in MMC

Traditional construction contracts treat time in familiar ways: milestones are set, delay events may trigger extensions of time, and liquidated damages apply if completion slips. Standard forms such as JCT and NEC assume a live site with visible progress and multiple overlapping packages. In that world, time is flexible and disruption is mostly managed on site.

MMC changes this logic. The programme now hinges on factory production slots, transport windows, inspection gates, and tight interface tolerances. Delays can occur not only on site, but also in the factory, the yard, on the road, or during pre-installation quality checks. Contracts must recognise this expanded time risk and allocate responsibility clearly. Without this, disputes escalate quickly and costs follow. This topic connects directly to procurement choices in Chapter 3 and supply chain structures in Chapter 4.

Factory Time Is Real Time

In MMC, the factory sits on the critical path. Production calendars are planned weeks in advance. Labour is booked, materials ordered, and inspection windows scheduled. If design is not ready, the factory slot is lost and often cannot be recovered within the same month.

Contracts should therefore reflect factory activities as time-critical in the programme. Clear rules are needed on:

- When manufacturing starts
- What counts as measurable progress
- What constitutes delay within the factory

NEC contracts can support this with early warning notices and compensation events; JCT contracts can be amended similarly. Both require explicit definitions to cover off-site activities and inspection gates. This links directly to the design freeze and gate logic in Section 6.3 and to the quality gates in Chapter 5.

Payment Linked to Time Milestones

Much of MMC's value is created before components reach site. For example, a bathroom pod may be nearly complete in the factory while the site shows little visible progress. If payments are only linked to site milestones, suppliers face significant cash flow risks, which can slow production.

A better approach is to include factory milestones as payment triggers. Evidence could be factory inspections, dated photographs, or digital progress records. When a pod passes a defined inspection gate, a corresponding payment percentage becomes due. This keeps the production line moving and reduces schedule risk. It also supports the golden thread evidence approach described in Chapter 5.

Programme Float and Interface Risk

MMC requires precision sequencing. Pods cannot be stored indefinitely. Panels must be installed in order. Deliveries are scheduled tightly to site readiness. While no project is perfect, traditional float hidden by resequencing is limited in MMC.

Buffers should be placed where risks occur, and must be visible and owned. Examples include:

- Buffer before design release to protect factory schedules
- Buffer around delivery to cover road transport risks
- Buffer around lifting to protect against crane availability issues

Contracts should name buffer owners and require approval before buffers are consumed. For example, if slab casting delays but pods arrive as planned, contracts must specify who covers storage, redelivery, and crane rebooking costs. This is both a time and cost issue, laying groundwork for Chapter 7.

Interface Delay Events

MMC relies on tight interfaces: pod to structure, panel to frame, cassette to core. If receiving structure is out of tolerance, installation cannot proceed, halting the entire chain.

Contracts should require:

- Site readiness checks
- Survey evidence
- Formal sign-off protocols before delivery

Tolerances must align with manufacturer guides. Responsibilities should be clearly assigned, e.g.:

- Frame contractor confirms levels and fixings
- MEP contractor signs off installations pre-dispatch
- Cladding contractor coordinates install and logistics plans

Clear duties reduce delays and link directly to coordination processes in Chapter 4 and inspection plans in Chapter 5.

Re-work and Remanufacture Delays

Chapter 5 highlights the cost of non-conformance; the time penalty is even greater. A non-compliant unit may need return to factory, reinspection, rebooking of transport, and waiting for a new crane slot. This can add several weeks per item.

Contracts should clarify:

- Whether delay is compensable
- Who pays for redelivery and reinspection
- How the programme adjusts if replacements arrive in different batches

Better still, contracts should encourage early detection via sample units, first-off reviews, mock-ups, and hold points. This protects time and cost simultaneously.

Supplier Failure and Continuity Planning

The MMC supply chain is still evolving. Even standardised modules often require project-specific tailoring. If a key supplier fails mid-programme, a simple swap is rarely possible. There is seldom an off-the-shelf equivalent that fits existing interfaces, risking severe time disruption.

Contracts and procurement plans should include continuity measures such as:

- Step-in and novation rights for alternate parties
- Escrow of design data and manufacturing instructions
- Clear intellectual property rights enabling client reproduction
- Compatibility standards for critical interfaces to onboard second sources quickly
- Parent company guarantees, performance bonds, and insurance
- A concise continuity plan detailing mobilisation timeframes for replacements

These are time control measures that belong alongside payment and delay clauses (see Chapter 3).

Storage, Just in Time, and Practical Realism

Ideal MMC delivery would be just-in-time: components arrive exactly when needed. In practice, current UK transport and supply chain maturity make this difficult.

Some contractors choose short-term storage near site to protect the programme. Storage adds cost and risk of damage but provides a sensible buffer against logistics uncertainty.

Contracts should specify:

- Who pays for storage
- Who insures stored goods
- How long items can be stored without risking damage or warranty issues

Over time, improved logistics and tracking should reduce this need. Meanwhile, small visible buffers are preferable to hidden risks.

Practical Clauses That Make Time Work

In summary, a short set of contractual clauses can have outsized impact:

- Factory time is on the critical path and included in the programme
- Design release dates and inspection gates are named and owned
- Payment milestones include off-site progress with clear evidence requirements
- Buffers are placed before design release, dispatch, and lifting, and have named owners
- Interface readiness checks and tolerance surveys are mandatory before delivery
- First-off approvals and mock-ups are encouraged and contractually allowed
- Changes after design freeze follow a fast impact review and defined approval route
- Continuity planning is documented, including access to designs and rights to complete

When contracts mirror the realities of MMC delivery, programmes become stable, suppliers can plan confidently, and disputes reduce. This supports supply chain discipline (Chapter 4), builds on quality gates (Chapter 5), and creates a strong platform for cost management (Chapter 7).

Time Management Tools and Planning Techniques for MMC Projects

Effective time management in MMC is not just about shorter bars on a programme chart, it requires a fundamentally different way of organising how time is planned, monitored, and communicated. Traditional planning often assumes a linear sequence of site work, but MMC involves many parallel activities, with key milestones in the factory as well as on site. Delays in one step can stall multiple linked steps. A good plan makes these links visible and manageable.

Building with Time Logic, Not Only Time Targets

Projects often start with a fixed handover date and work backward, which is necessary for funding and reporting but insufficient for MMC. The plan must clearly show how each task depends on others, spanning factory and site. For example, pods can't ship until drawings are frozen and a dry landing zone exists; panels can't be installed until the core and fixings are ready; services can't be tested until factory and site parts are connected and powered. Visible dependencies prevent guesswork and help manage the design freeze, supply chain inputs, and inspection gates effectively.

Digital Sequencing and the Value of 4D

A simple 4D review links the programme to a basic model and animates the build sequence. This is valuable in MMC where finished elements arrive on a tight site with limited crane access. 4D helps identify clashes, crane reach issues, or conflicts in landing zones. The goal is shared understanding, not a polished video, reducing late changes and making time more predictable, supporting subsequent cost management.

Factory Calendars and Live Supply Chain Input

Factory production runs on strict calendars, with labour, materials, and maintenance assigned to specific weeks. Unlike site work, factories can't easily add shifts at short notice. The site plan must respect this. Agree a joint calendar with the manufacturer including design release dates, first off inspections, quality gates, and dispatch windows. Treat these as firm constraints and meet weekly to update risks and delays. Keeping the supplier inside the planning loop makes the programme realistic; outside the loop, it's just hope.

Gateways and Hold Points That Protect Time

Fast programmes fail when control is weak. Gates and hold points, brief, clearly owned checkpoints like design freeze approvals, factory first-offs, pre-dispatch quality checks, site readiness sign-offs, and interface verifications, are crucial. They prevent wasted effort and enable confident progression. Skipping gates may save time upfront but usually causes longer delays. Treat these gates as critical tasks.

Buffers That Match Real Risk

MMC shifts uncertainty rather than removing it. Instead of a large hidden float at the end, place small, visible buffers where risk actually lies: before design release, within production batches, around deliveries, lifting, and testing. Each buffer should have an owner and be reviewed weekly to avoid being eaten away. Where transport reliability is low, short-term site storage may be used as a controlled buffer, with rules for inspection, insurance, and duration. The long-term goal remains smooth flow with minimal handling.

Master Plan, Micro Plans, and Look Aheads

MMC requires three levels of time control:

- *Master programme*: Sets overall strategy and key dates
- *Micro plans*: Control specific streams like pod installation or facade panels
- *Look aheads*: Focus on the next three to four weeks of live factory and site work

All levels must stay aligned. Weekly meetings help update actual vs planned dates, review risks, confirm upcoming gates and responsibilities, and check delivery bookings against crane and crew capacity. A shared tracker with named owners keeps the process transparent and effective.

A Joined-Up Way of Working

When plans show clear logic across factory and site, when sequences are tested with 4D, when factory calendars are integrated, gates respected, and buffers well placed, time becomes predictable. This strengthens the quality system (Chapter 5) and supports cost management (Chapter 7). MMC then delivers calm, controlled speed instead of frantic uncertainty.

The Speed Effect: Benefits and Pitfalls of Faster Delivery

MMC's promise is speed – faster delivery, faster market access, faster handover. This promise can be fulfilled, but only if the whole system is designed and managed accordingly. There's a clear difference between projects where speed is a real advantage and those where it exposes lack of readiness.

When Faster Is Actually Better

Some sectors benefit strategically from speed: hospitals adding capacity before winter, schools expanding during holidays, retail shortening downtime during refits. MMC allows factory work to run in parallel with groundwork, delivering completed rooms quickly and reducing weather impacts. The result: shorter site occupation and earlier building use.

The Risks of Rushing

Speed alone is no plan. Fast installation introduces new risks if sequencing is weak. Interface problems, like mismatched fixings or tolerance issues, can stop installation. City sites risk delivery congestion without phased bookings. Compressed checks risk quality failures that surface post-install, causing longer fixes. These problems arise when installation outpaces design, logistics, and quality control, not because MMC itself is fragile.

Local Authority Constraints

MMC projects often operate under strict urban controls on working hours, access, noise, and road closures. Speed only helps if these controls are booked early and built into the plan. Large modules need road closures weeks in advance; crane time is limited to certain windows. Ignoring these rules makes speed gains theoretical.

Hype and Reality

Marketing often highlights headline figures like "30% faster" or "half the site time". While possible, these claims rarely include time spent before site starts, coordination, procurement, mock-ups, logistics, stakeholder approvals, and design freeze. Early stages often take longer for MMC newcomers. Time saved on site is real, but overall timelines shift. MMC should be framed as a different timeline, not a magic one, moving effort earlier for calmer later stages.

Public Perception and Client Confidence

In the UK, speed can be perceived as poor quality. Fast build equals flimsy, in public opinion, which affects market reception and funder confidence. As Chapter 5 showed, factory production can improve quality when well controlled. Similarly, speed in MMC comes from control, not cutting corners. Projects sharing clear evidence of quality, third-party assurance, and performance help shift perception. Framing speed as a product of planning and process builds trust rather than eroding it.

The Design, Make, Install Disconnect

A common cause of hidden delay is poor integration of design, manufacturing, and site teams: design sign-off may come too late for factory schedules; factory batches finish before site is ready; install teams arrive without final logistics data; post-install testing reveals mismatches. Delay in MMC rarely appears as single days, it comes in cycles. Missing a factory or transport slot forces complex recalibration.

A Balanced View

MMC offers speed when discipline is in place: early decisions, clear logistics, firm gates, and regular communication. When present, projects proceed steadily and finish early. When missing, faster methods simply accelerate risk exposure. The goal is reliable time that protects quality and cost, not speed for its own sake. The following sections explore how staff knowledge, daily control, and commercial structures make faster delivery also better delivery.

The Human Factor: Staff Knowledge, Training, and Readiness

Success in MMC depends not just on technology but on people's understanding and coordination. Time control requires that everyone, from planners to site crews, knows what must happen, when, and how to respond if something is delayed. The greatest time savings come from removing sequential site tasks, replaced by factory-built components arriving ready to install. However, this only works when the team fully grasps the new workflows and control points.

- System understanding as the base for speed: Time in MMC is a linked chain where each step depends on the previous one. Everyone must understand their role and the dependencies to avoid delays that ripple through the process. Training shifts focus towards planning, coordination, and verification, with time delays often visible immediately at control gates.
- What the site manager needs to do differently: Site managers become choreographers and controllers, managing deliveries, lifts, and handovers while ensuring readiness and refusing installation without clear approvals. They need skills to read inspection reports, understand lift plans, and run site readiness checks. Short, focused training significantly reduces delays.
- What the factory manager needs to change: Factory management must synchronize production with site readiness, ensuring batches arrive in correct sequence and timing. Live communication and shared calendars between factory and site create a partnership rather than a supplier relationship, avoiding idle inventory or delays.
- How to train for real projects: Effective training is project-specific, using real interfaces, lifting paths, and tolerances. Early involvement of subcontractors, mock-ups, digital walk-throughs, and cross-shadowing between factory and site improve understanding and reduce risks.
- Communication that saves days: Clear communication prevents costly errors like mis booked deliveries or unsigned inspections. Daily short huddles with all key parties focus on upcoming tasks, risks, and immediate decisions, maintaining clarity and avoiding delays.
- Competence and certification: MMC requires new skills, verifying interfaces, reading inspection records, and using digital tools. A competence framework defining necessary skills and certifications ensures the right person performs the right check at the right time, improving reliability.
- Crew rhythm and standard work: Repetition and steady rhythms improve speed and reduce mistakes. Consistent tasks, fixings, and sequences build crew confidence and efficiency, saving time without added stress.
- Readiness checks and authority to stop: Quick readiness checks before lifts confirm everything is ready. Workers must have the authority to stop work if conditions aren't met, preventing delays caused by re-work or unsafe operations.

- Simple measures that predict time health: Track a few straightforward metrics, plan reliability, first-time install rate, crane utilization, inspection pass rate, and information cycle time, to monitor schedule health and drive continuous improvement.
- Leadership, culture, and public trust: Leadership enforces discipline at gates, values inspection time, and encourages early problem reporting. Sharing transparent quality evidence helps overcome public scepticism that fast means poor quality, building confidence among clients and users.
- Bringing it all together: MMC success hinges on the team's capability and coordination. Daily communication, respect for gates, targeted training, mock-ups, and simple tracking produce a steady, predictable programme with fewer surprises, disputes, and delays, delivering real, repeatable time benefits.

Linking Time to Cost: The Value of Predictability

In construction, time and cost are deeply connected but often difficult to manage effectively. Traditional approaches tend to see faster delivery as more expensive due to overtime, premium logistics, and resource juggling. Modern Methods of Construction (MMC) shift this dynamic: speed alone is not the main value, but predictability in time leads to efficiency and cost savings by reducing uncertainty.

A project that takes longer but follows a disciplined, predictable plan often outperforms a rushed project plagued by delays and re-work. Predictability is MMC's key currency, improving both schedule and budget in measurable ways.

- The real cost of programme drift: Delays cause direct costs (idle cranes, turned-away deliveries, waiting crews) and indirect costs like double handling, extra logistics, and loss of crew rhythm. These small inefficiencies erode margins, damage trust, and increase disputes. Quality gates, introduced earlier, protect both time and cost by preventing costly downstream fixes.
- Early spend as a way to lower total cost: MMC often requires earlier financial commitment, locking designs, securing factory slots, ordering long-lead items. Though it frontloads spending, this early investment avoids premium charges, stabilizes supply chains, and reduces risk premiums, ultimately lowering total cost. Contracts that tie payments to off-site milestones support this approach.
- Repeatable work reduces risk: Projects with a high proportion of off-site manufactured work (pre-manufactured value) benefit from repeatable, standard components. This stabilizes installation times, reduces defects, improves first-time install rates, and lowers contingency allowances. Consistent tasks speed crews up and reduce overhead and change costs.
- Clear time logic lowers tender risk: When programmes clearly define interfaces, factory dates, and milestones, suppliers reduce risk allowances, leading to sharper bids, better terms, and more accurate cash flow forecasts. This clarity also smooths commercial management, reduces disputes, and reassures funders by linking payments to tangible progress.
- Lead in time and the shape of funding: MMC's earlier spend profile requires funders to adjust drawdown schedules to match design releases, factory gates, and dispatch windows. Aligning finance with programme needs avoids cash flow stress, prevents suppliers from carrying unsustainable costs, and maintains trust and schedule integrity.
- Measuring predictability in simple ways: A few straightforward metrics, percent plan complete, first-time install rate, crane utilization, inspection pass rate, and information cycle time,

reliably indicate project predictability. Tracking these weekly provides insight into schedule health and directly relates to cost control, without complex dashboards.
- Buffers that cost less than delay: Buffers are not waste but strategic protections. Small buffers before design release, within batches, or in delivery windows prevent costly disruptions like missed production slots or lost crane time. Controlled storage buffers, with clear contractual terms, mitigate logistics risks and reduce expensive re-work or out-of-sequence deliveries.
- How time discipline becomes cost discipline: Time and cost discipline are two sides of the same coin. Tools like design freezes, quality gates, factory calendars, and weekly reviews enforce both. Procurement, supply chain, and quality management all contribute to time control as a means of cost prevention, not cutting costs but avoiding waste.

MMC does not promise to be cheap or fast by default. Instead, it offers a disciplined way to reduce uncertainty. Predictable schedules lead to fewer claims, less wasted effort, minimized idle time, clearer valuations, and fewer disputes. A nine-month project with steady progress typically costs less than a rushed six-month project battling delays.

The next chapter will explore cost management, showing how predictable time translates into measurable savings and how commercial structures can support MMC's component-based delivery without obscuring costs. Time and cost must be managed together for success.

Summary: Reframing Time in MMC as Design, not Delivery

Modern Methods of Construction (MMC) can shorten on-site work, but speed is not automatic, it depends on early design decisions and disciplined management. Time savings come from complete information, aligned procurement with factory capacity, and integrated factory-site planning. Simply put, MMC achieves faster builds by working earlier and assigning clear responsibilities upfront.

Time in MMC is front-loaded. Successful site programmes flow smoothly because complex tasks are resolved beforehand. Firm design freezes, respected gateways, and logistics aligned with installation windows make delivery seamless. Failures in any of these areas cause delays and reduce confidence.

MMC is less forgiving of late changes compared to traditional construction. Early or late arrivals disrupt multiple trades, and missed interfaces can halt both factory dispatch and site installation. Thus, time should be managed as a design outcome, not an afterthought. Choices in components, design maturity, procurement, and contracts must all align from the start to control time effectively.

The greatest advantage of MMC time management is predictability. Reliable schedules reduce contingency buffers, increase funder confidence, enable client planning, and help supply chains optimize labour and transport. This foundation enhances cost control and quality, reinforcing lessons from earlier chapters and setting the stage for the cost management focus in Chapter 7.

Strong time control benefits extend beyond construction. Earlier completion enables faster occupancy, smoother commissioning, and better operational starts. When time is stable, cost and performance naturally improve.

Chapter 7

Cost Management

Introduction

This chapter examines the cost implications of employing Modern Methods of Construction (MMC) within the construction industry, showing how utilising the key themes discussed earlier in the book, can reduce overall cost on a project. While earlier chapters have introduced aspects of cost in relation to specific themes, this section provides a more comprehensive exploration. Drawing on findings from Van Vuuren and Middleton (2020), the discussion will consider the principal cost variables within construction and assess how these may be reduced, eliminated, or remain unaffected by the adoption of MMC.

Given the centrality of cost to construction projects, this topic warrants detailed analysis. The chapter will therefore connect the cost dimension not only to wider economic debates but also to issues of productivity and efficiency. Particular attention will be given to individual cost drivers, including the influence of weather, project duration, and re-work requirements.

The discussion will further evaluate how MMC alters traditional cost management practices. This includes consideration of earlier payment regimes, the potential for accelerated income streams, and the financial implications of procuring materials and systems prior to their arrival on site, representing a departure from conventional approaches. Two headline impacts are worth noting at the outset: the tendency for MMC to bring design, procurement and factory spend forward while shortening site duration and thereby reducing preliminaries; and the role of embedded quality assurance gates in preventing defects rather than correcting them, with implications for re-work, claims and programme certainty.

Finally, the chapter will recognise that the extent of cost benefits is closely linked to supply chain maturity. As highlighted in Chapter 4, supplier consolidation and stable pipelines play a crucial role in enabling predictable pricing and reliable lead times, thereby shaping the commercial models through which MMC can deliver value.

Construction Economics and the Impact of MMC

Construction economics explores the transfer of wealth and the processes for doing this. There are many aspects to consider when looking at construction economics and some of these will be affected by a change in production type to off-site construction. These are discussed and evaluated in more detail below.

DOI: 10.1201/9781003570431-7

Construction Sectors

The construction industry is often one of the biggest industries in most countries in Europe the rate generally ranges between 4 and 6% of GDP – while other sectors such as financial services or agriculture (depending on the country) may be larger, the construction sector is often seen as the industry that gives the clearest indication of economic activity. In other words, if the construction sector slows it is often a key indicator of an economic recession looming. Often that is one of the reasons that governments of the world wanting to drive economic growth will invest in infrastructure projects.

The industry is made up of a number of sectors and sizes of organisations, also making it one of the most diverse industries to operate in. In 2024 the UK sectors was made up of 337,000 (92%) companies up to eight employees, 22,000 (6%) up to 25 employees, 4,000 (1%) up to 80, 1,000 (>1%) up to 600 employees and 100(>1%) over 600 employees (ONS 2024).

The sectors can often be split into a number of different ways, but a typical way of splitting sectors is by client type and could be listed as:

- Central Government
- Local government
- Residential and Domestic
- Commercial
- Retail
- Leisure
- Education
- Health
- Other

Each of these sectors will have varying consideration for off-site use, and they have been explored in more detail below with some case studies to support.

Central Government

Typical projects commissioned by central government include civil infrastructure works, schools, and prisons. Increasingly, these projects are turning to off-site techniques, with elements such as bridges and tunnel sections manufactured in factory settings and then delivered to site. Similarly, new classroom blocks have been delivered using modular construction, while in the prison sector entire cells can now be prefabricated off-site and installed as complete units.

From a cost perspective, central government is a particularly influential stakeholder. The promotion of Pre-Manufactured Value (PMV) targets is intended to encourage pipeline standardisation, thereby enabling steadier factory pricing and supporting long-term investment in the supply chain. This links directly to the themes of supplier consolidation and predictable pipelines discussed in Chapter 4.

Local Government

Projects led by local government tend to focus on the repair and maintenance of infrastructure originally installed through central government programmes. As such, there is generally limited scope for off-site techniques. From a cost perspective, local government projects are significant in terms of lifecycle savings, as even modest reductions in maintenance time or disruption can yield measurable financial benefits.

Residential and Domestic

Residential projects encompass both new-build and refurbishment of domestic properties, ranging from low-rise housing to high-rise apartments. The repetitive nature of many residential designs makes them well suited to off-site construction. However, uptake remains relatively low in many areas, influenced by public perception and profitability considerations. Cost-wise, the potential for reduced construction time and standardised component production can improve cash flow and decrease labour costs.

Commercial

Commercial projects typically include factories, warehouses, and large-scale industrial facilities. While structural components such as beams and columns can be manufactured off-site, the expansive floorplates often make full panelised or modular systems impractical. In cost terms, off-site techniques can still reduce programme delays and material waste, providing targeted financial efficiency rather than wholesale savings.

Retail

Retail developments similarly favour large, open-plan spaces, which can constrain the application of off-site methods. Nonetheless, modular or prefabricated elements can still accelerate construction and reduce on-site labour costs where suitable.

Leisure

Leisure buildings include hotels, public houses, and sports centres. Hotels, with their repetitive room layouts, are particularly amenable to modular construction, a method already in use within the industry. Even where open spaces are required, partial off-site solutions can deliver cost benefits through programme acceleration and reduced site disruption.

Education and Health

Educational and healthcare facilities, such as schools and hospitals, often follow standardised designs that lend themselves to off-site techniques. Cost advantages include reduced site duration, lower preliminaries, and fewer disruptions to existing operations, supporting both financial and operational efficiency.

Accounting and Finance

Accounting and finance in construction relate to the processes and methods used to manage the finances of all stakeholders involved in a project. Effective financial management requires an understanding of the various contributors to project expenditure, ranging from straightforward factors such as the client, main contractor, and subcontractors to more complex stakeholders including suppliers, financiers, and planning authorities. The roles of these stakeholders, and their financial interactions, are considered in more detail below.

Client

The primary responsibility of the client is to fund the project, typically through a combination of reserves, investment, or borrowing to ensure that construction progresses according to programme. The adoption of off-site construction can alter both the timing and method of funding. While faster project completion can enable earlier income, this often requires earlier cash drawdown to cover procurement and factory spend. Financiers may also be less familiar with non-traditional approaches and may require additional assurances before committing capital. Consequently, clients must adapt their financing strategy to balance cashflow requirements and maximise the net present value (NPV) benefits associated with shorter project durations, as discussed in Chapter 6.

Main Contractor

The main contractor is responsible for delivering the construction works while ensuring profitability. Financial management practices have a significant effect on the contractor's ability to achieve this. Profit margins in the sector are typically low, often ranging from 1–5%, with an average around 2%. Within this context, the adoption of MMC can have substantial financial implications. For example, the need to pay for materials and factory-fabricated components earlier in the process can strain cash flow, while the reduced flexibility to influence production processes, particularly when a single supplier delivers a modular solution, may limit opportunities to maximise profitability. Conversely, benefits such as shorter project durations, lower health and safety costs, and reduced on-site overheads can improve financial outcomes.

Subcontractor

Subcontractors undertake work that the main contractor cannot, or prefers not, to perform themselves. As with main contractors, their profit margins are typically low. MMC projects can alter subcontractor dynamics, particularly in modular builds where a single supplier may deliver a substantial portion of the works. This can reduce the number of subcontractors required, with mixed financial consequences: fewer subcontractors simplify project control and reduce administrative costs, but reduced competition may limit cost-saving opportunities for the main contractor.

Suppliers

The role of suppliers in MMC projects is more complex than in traditional on-site construction, due to the additional tier introduced into the supply chain. In conventional projects, materials are delivered directly to site and installed in situ. In contrast, MMC requires materials to be delivered to a factory, manufactured into components or modules, and then transported to site. This additional step shifts a degree of cost control from the main contractor or client to the supplier.

Where multiple suppliers are involved, purchasing power can be leveraged to reduce costs. However, in modular projects where a single supplier provides the majority of components, this reduced competition can limit financial incentives for the contractor and client to adopt MMC. Panelised systems present a slightly different scenario: only a few key elements (walls, floors, roofs, etc.) are supplied in panels, leaving a larger proportion of the project open to competitive procurement, which can help manage costs and maximise profits. Nonetheless, from a contractor's perspective, this limitation is a notable drawback of MMC.

From the supplier's point of view, the primary financial consideration is the level of deposit required from the contractor. Large-scale, bespoke components, often produced for nationwide projects, carry higher risk than standard off-the-shelf materials. Even if full payment is received before dispatch, the supplier may incur costs if the contractor reneges on the agreement, necessitating storage or re-working the product for alternative sale. By contrast, standard items such as doors can generally be redirected to other projects with minimal loss.

Consequently, suppliers generally prefer to secure full payment before commencing production, which can be many months in advance of delivery or site commencement. Clients and contractors, however, are often reluctant to pay the full amount upfront. In practice, a negotiated deposit is typical, balancing the financial risk for the supplier with the cash-flow constraints of the contractor and client.

Funding Mechanisms

Construction projects rely on a variety of funding mechanisms, each with distinct implications for project delivery and cost management. While a more detailed discussion of funding impacts is provided later in this chapter, it is useful to. The three principal mechanisms for financing construction projects are:

Investment Loans

Investment loans involve an organisation, such as a bank, building society, business angel, or other investor, providing funds to support the project in return for an expected financial return. The terms of these loans can vary significantly. Established banks may offer lower, fixed-interest loans, with repayment obligations independent of project performance. By contrast, investors may demand higher returns linked to the success of the project, effectively sharing in the financial risk and reward.

This distinction has implications for the adoption of MMC. Banks are generally less concerned with project type or expected profit and therefore exert minimal influence over decisions regarding production methods. Investors, however, may perceive off-site construction, such as modular builds, as higher risk and could be reluctant to provide funding, potentially constraining the client's production choices.

Reserve Funding

Reserve funding utilises an organisation's own financial reserves to support project delivery. While this approach avoids interest payments associated with borrowing, it carries opportunity costs: funds used for construction are unavailable for other investments that could generate returns. Depleting reserves can also weaken the organisation's financial position, reducing liquidity to meet short-term liabilities and potentially affecting accounting metrics.

Off-site construction can mitigate some of these concerns due to generally shorter build durations. By accelerating completion, MMC can shorten the period between investment and return, reducing the time the organisation operates in a financially constrained state.

Pre-Selling

Pre-selling, or selling "off-plan" involves customers purchasing all or part of a project before completion, often at a discounted rate. This approach is commonly used in residential housing and, in some cases, large-scale venues such as football stadia, where season ticket holders or executive box purchasers pay in advance.

While pre-selling can provide early cash flow, it relies on customer confidence in the project's delivery. Changes to production methods, particularly the adoption of modular or off-site construction, may introduce perceived risk, potentially deterring pre-buyers. Therefore, careful communication and demonstration of project reliability are essential to secure advance funding through pre-sales.

Planning Authorities

Planning authorities are responsible for ensuring that developments align with the objectives of the Local Development Plan (LDP). The LDP sets out permissible developments within a given area and is overseen nationally by the Secretary of State, providing a country-wide strategic perspective.

While local authorities may not have a preference regarding construction methods, the timing of delivery is critical. LDPs specify sectoral outputs over defined periods, for example, 10,000 homes, one hospital, and ten hotels over ten years. If production is skewed towards one sector, it can disrupt local economic balance. For instance, a rapid delivery of hotels before residential housing could result in a shortage of local workers, whereas building homes first without corresponding industrial development may create unoccupied properties and economic inefficiency.

Shorter site durations offered by off-site construction can provide additional financial benefits from a planning perspective. Reduced construction periods can lower traffic management costs and minimise road possession fees, contributing to overall reductions in project preliminaries, as discussed in Chapter 6. Consequently, programme compression through MMC not only accelerates delivery but also delivers cost efficiencies that are directly relevant to planning authorities.

Local Communities and Other Local Government Bodies

Local communities, including residents, businesses, and other municipal bodies, have a vested interest in nearby construction projects, as developments can produce both positive and negative impacts. New housing or commercial infrastructure can stimulate local economies through increased spending, but it can also place pressure on services such as schools, healthcare, and public transport.

Public perception plays a critical role in project success. The rapid delivery afforded by off-site construction may be viewed negatively if perceived as compromising quality. For example, a housing development that traditionally takes five years but is completed in one year might face scrutiny, potentially affecting property values and community acceptance.

Off-site construction can also reduce local disruption. Consolidated modular deliveries mean fewer trips to site, lowering transport-related disruption and associated costs for the community. This links directly to the logistics considerations discussed in Chapter 4.

To address potential concerns, stakeholder engagement and education programmes are often necessary. Raising awareness of the benefits of off-site construction, and clarifying differences from traditional on-site methods, can help manage public expectations, maintain confidence in the local market, and mitigate potential social or economic disruptions.

Economic Terms and Considerations

As well as the stakeholders listed above that may be affected by the production choice, some of the key economic principles such as supply and demand, law of diminishing returns and opportunity cost can also be affected by the production choice.

Supply and Demand

Supply is the amount of a product that is provided to the marketplace, and demand is the amount of need or want of that product. The ideal situation is called equilibrium, which is where the amount of supply matches the amount of demand. Supply is seen to be generally within the control of the producer, whereas demand can be influenced by the producer but is generally in the control of the buyer.

Supply will depend upon the amount of productivity the producer can achieve, and the amount of production will depend on the amount of investment put in. For example, in a construction company the more the company invests in labour and updated equipment, or in faster productivity methods (Such as off-site construction), the quicker the product will get to market. The more profit that a company will make from a product, the more they are likely to invest in that product to get it to market, and vice versa. Therefore, supply is very often a response to demand and often successful business are the ones who can predict most accurately the movements in demand and prepare for that eventuality. It can also be assumed, this may also be a reason why new products to market and new business can often fail in the first year or two, as the demand has not had chance to grow alongside the supply.

Demand is mainly led by need and want, products with a need are easier to understand as they tend to feed a sector that has to have the product and fulfils a basic requirement of human nature. This can be staple food products, shelter, health, clean drinking water, etc. In Maslow's hierarchy of need, the bottom level or maybe the bottom two, tend to fit in with this category. With items that are in need, it is rarely influenced by marketing input. People will tend to buy what can be afforded with the basic need items.

Products that fall into the want category, are more led by desire than a must have, and these can be influenced by marketing. It moves beyond the basic human requirement into the realms of a nice to have and a luxury item. Depending on the upbrings, culture, country, and social outlook of the human buying the products, what is classed as a basic need and luxury item can change. For example, a young adult in some more affluent countries may see owning a car as a basic need, where in some third world countries it is very much a luxury item. Marketing can take many forms and can include word of mouth, media outlets (advertising and product placement) including newspapers, television, social media, etc., celebrity endorsement, peer pressure, society pressure, etc. and these can all also be altered depending on where you live. As has been stated above one of the key barriers to the uptake of off-site construction techniques is public perception. In certain cultures, and countrles, anything that is build or produced quicker than expected, or than is traditional, is perceived to be of lower quality. And it can take a long time for public opinion to change. An examination of key engineering processes in the 20th century shows that, for example, the motor car industry used to have all vehicles built by hand, and the introduction of mechanised and robotic processes and production lines was met with scepticism.

However, nowadays hand built motor cars are now given that same scepticism with regards to quality standards as the general public expects robotic built cars to be the highest quality. The housing industry is currently, going through the same transition, and until off-site construction products are seen as having the higher quality output, than the hand built traditional products, then the demand and therefore supply will remain low. The assumption from those that work in the sector, is that factory produced products will have greater levels of quality as they are not subjected as much to the impact of weather, and human errors, lack of control, etc., than can happen on site. As yet, not a lot of off-site construction processes have reached the levels of automation that can be seen in other engineering sectors, but even with the increased human power involved, the greater control and supervision levels and standardised processes, leads to greater quality standards. Once this is understood by the population at large, plus the benefit of a few endorsements from government and celebrities, then the demand will increase and that will in turn lead to greater supply.

A further consideration that impacts on demand is the price of the product – the cheaper it is the more the demand (but also the reduced profit and therefore potentially the reduced supply). Industry is trying to achieve the best product at the lowest cost, but highest price that they can achieve within the given marketplace. This will lead to the greatest return on investment. If industry can lower the cost and in turn lower the price but maintain quality and profit margins, then there is likely to be increased demand for the product.

Equilibrium is achieved when the supply and demand meet. This occurs when the producer has got the price, quality and marketing of their product to exactly match the amount of supply they can produce. The difficulty for industry is that the demand, need, cost, etc. is a moving target, so the marketplace is nearly always in what is sometimes termed a "buyer's market" or "seller's market". As an example of this let us again turn to the UK low-rise single-family occupancy housing market. The housing market in the UK, as in much of the world, is in crisis – this is the term used when there is insufficient supply to meet the demand, and governments around the world are constantly trying to find ways to reduce this deficit. When there is a seller's market, that is when there is more demand than supply, the cost of items will increase, and profit margins remain high. This is the current state of the housing market in the UK. If the housing market demand reduces, either by the demand for housing going down (a dwindling population, for example) or by the supply increasing (house builders building more), then the costs will reduce and the profit margins accordingly. As has been shown above, several stakeholders can influence this – governments can help by increasing the planning permission allocation, easing land use restrictions, and height restrictions, etc., so that more residential properties can be built. Local communities and the public at large can help to influence this by demanding products that can be built quicker – for example, this could be more high-rise properties, etc. – and the developer themselves can help by building faster, although this may not be possible, as if the developer releases more houses into the market, then profits will drop.

Economy of Scale

One of the fundamental economic principles is that if you can produce a product in sufficient quantities, it will become cheaper to produce and therefore cheaper to market, thereby increasing the demand. Therefore, the principle of using more MMC products will increase the benefit – the more it is used the more financially beneficial it becomes. One of the key reasons, however, that more MMC is not used is often stated as the additional cost, which is explored more below. It is one of the phenomena of economics that cost prevents the use of a product, but the more it is used the cheaper it becomes.

Law of Diminishing Returns

The law of diminishing returns states that when efficiencies are introduced, they can pay big dividends, and produce good returns. However, as more and more of the efficiencies are introduced the amount of return reduces, and eventually this can also reverse and cause a loss as opposed to a benefit.

Opportunity Cost

Opportunity cost is the principle that if you choose one production method over another then benefits of choosing one, becomes a cost on the others – as in a missed opportunity to gain the benefits by forgoing the other method. An example of this could be that when using a traditional production on-site method one of the key benefits is gaining the input from skilled workforce providing a flexible approach to the production. If off-site production is used, then this benefit is not available (or needed) but the principle of opportunity cost places this loss in benefit as a cost against the selected off-site method.

Economic growth

World economies generally grow from year to year – in times of high growth they move forward faster, and in times of recession growth slows and sometimes stalls, but generally the economy today is stronger than it was ten years ago. As the economy grows the cost of items grows, products cost more, peoples salaries grow to match (hopefully) and when looking at the costs involved in construction projects that can last for a number of years, this growth needs to be taken into account. The typical process used is by linking cost analysis with Net Present Value (NPV) that brings all future costs back to an equivalent cost in today's economy. More on this process is discussed below.

Productivity

The key aspect of the country's economy is the level of productivity – when productivity grows generally the economy grows, finding more efficient ways of working aids in increasing that productivity, so there is typically a correlation high innovation, high productivity and high economic growth. One of the reasons is that countries that invest in education are also countries that perform well economically. The construction industry has had stagnated productivity in the UK for a number of years with only minimal growth compared to other sectors. One of the key reasons for this is that while there is lots of research into innovation in the sector developing systems of productivity efficiencies such as BIM, Lean, MMC, etc., these innovations are often proven to provide benefits that outweigh the drawbacks, The industry is often reluctant to take up these processes. Often not considering it is better to ensure the profits being made currently are not jeopardised by a risky research and development programme to possibly provide more benefit in the future. This risk aversion is one of the reasons for slow growth in productivity. Government intervention does sometimes try to force the industry to innovate, such as when providing industry changing reports such as Latham Report, and Egan reports, or stepping in and mandating the use of an innovative change, such as happened with BIM in 2016. But if an industry is reluctant to change, they will often find a way to get around the uptake. As with many ventures like this, it is often brave pioneers who lead the way, some succeeding and some failing, until eventually the whole of the industry sees the benefits. This has happened many times in the past with the introduction of processes such as Auto CAD, trussed rafters, battery operated screw drivers, nail guns, etc. It takes a while for the industry to appreciate the benefits, but eventually the process becomes the norm, this will happen with BIM and the use of MMC although it may take a while for the industry to get there.

BIM has the potential for having a deep impact on the finances of a project, for much the same reasons as MMC. For example, the potential to cut out errors and improve quality with the complete three-dimensional design of all components and the ability to run through clash detection. There is the limitation of need for design changes and their financial impact, as the client has the ability to fully immerse themselves in a 3D environment when the project is still in the design phase and make all of the decisions needed prior to starting production. The inclusion of a fourth dimension of BIM that integrates time management process into the design so that a more accurate time certainty program can be developed. Equally with 5D for cost certainty, etc. The potential for standardizing processes and information when using BIM enhances the potential for financial savings when using MMC, if the BIM process is combined with the MCC product there is scope for many cost benefits.

Cash Flow Analysis

Cash is king, is often touted by economists – the principle being that if you have strong, readily available finances to ensure coverage of any short-term liabilities then your company will be strong and survive. One of the best tools to control and manage short-term liabilities and the short-term cash, is called a cash flow forecast. A cash flow is produced for projects and also an overarching one is produced for organisations. The key principle is to plot out categorised costs and income, using experience to provide the best educated guess possible, with the knowledge to hand. An often-forgotten process of a cash flow forecast is the need to continue it through the stages of cash flow management and cash flow analysis. Cash flow management is the process of recording actual costs and actual income as they become known, and for the forecast to be adjusted accordingly for future management decisions. The analysis section looks back over the predicted cash flow and the actual cashflow, to develop the knowledge and experience of those producing them to make future cashflows more accurate.

Cashflows, once developed, can also be used as decision-making tools, in order to make more profit for an organisation – for example, when comparing production techniques, a cash flow showing outcomes and income for a traditional project can be developed and compared to the cash flow for an MMC project. This will show how the shorter timescales will impact on overall financial management.

Value Engineering and Management

This is a term that is often interchangeable in academic research as well as in the industry, the most common term used is Value Engineering. The basic principle is that activities within a project are broken down into smaller parts, each part is analysed and the most efficient way of doing that activity. Once this has happened to all activities then the entire project should be the most efficient it can be. Value engineering can be an expensive process, especially on built environment projects which are unique, as such a balance is needed between the amount of depth that each activity is explored to, against what the potential return may be. In manufacturing, the returns are easier to ascertain as the processes are more repetitive and therefore costs saved on one process can be replicated and make savings over all of them. But as on-site construction is unique, no two projects are the same project in the same place, built by the same people, it is harder. This also helps to show that economy of scale as discussed above, will also pay more returns if VE is also used in the manufacturing processes.

Skills Shortage

One of the key issues facing the modern construction industry over the last few years, which appears to be getting more of an issue, is that of lack of skills in the correct area. This is a particular issue in trade skills and professional skills and knowledge, many of which are required for on-site production processes. Using off-site construction removes the need for the highly skilled trades on a construction site, therefore the more that off-site processes can be used to replace on-site processes, the less the skills shortage becomes an issue. From a profession specific activity consideration, it is more to do with knowledge and understanding, as the professions are being trained and developed to fulfil the gap, and the more that these personnel are aware of the MMC processes and their advantages and disadvantages, the more the industry will benefit. This is one of the key reasons for writing this book, and why you are reading it.

Cost Variable Analysis

Cost is quite elusive as a subject matter – there are so many aspects to the costing of any item and so many variables that impact on what the overall cost is, that it can get quite complicated, often leading to a balance being struck between the time that can be allocated to calculating costs against the time that is available to do the calculation. There are several sources that have tried to identify all of the cost variables that impact on a construction project and therefore need to be considered when making a decision on whether to use traditional or some form of MMC as the production method. Within this section there is a breakdown of each of the cost variables identified from several sources, although the key sources are a paper published by Van Vuuren and Middleton (2020) that had established cost variables that should be considered for off-site and on-site production and expressing the difficulty in monetising these variables, plus content from *Construction Quantity Surveying* (Towey, 2018).

Cost Considerations

There is a common perception within the construction industry that modular production techniques are more expensive than traditional construction methods, while panelised systems are considered comparable in cost. However, it is essential to contextualise this research within the evolving landscape of construction practices. The traditional construction methods examined in the 2007 National Audit Office (NAO) analysis differed significantly from those used today. At that time, factors such as air permeability were of minimal concern, fire safety regulations were less stringent, and energy consumption requirements were considerably lower.

In this context, "traditional construction" refers to site-based, labour-intensive building practices. Additionally, fluctuations in material costs influence cost comparisons between construction methods. For example, the price of copper, a key material in traditional plumbing systems, declined sharply before rising again following the COVID-19 pandemic. Modular construction, which typically requires less copper, is less affected by such price variations. Consequently, both material and labour cost fluctuations play a role in determining the relative expenses of modular, panelised, and traditional construction methods.

Order of Cost Estimates in Construction

The Royal Institution of Chartered Surveyors (RICS, 2021) has developed an order of cost estimate framework that categorises the key cost components of a construction project during the early stages of development. These categories are discussed in detail below to provide context for the cost elements. The primary cost categories include:

- Facilitating Works
- Building Works
- Preliminaries
- Overheads and Profit
- Project and Design Fees
- Risk
- Inflation
- Tax

Facilitating Works

Facilitating works refer to any preparatory activities required to make the construction site ready for development. These may include ground remediation, demolition of existing structures, and service diversions. Since these works are necessary regardless of the chosen construction method, they are not considered further in this analysis when comparing production types. It is important to note, however, that as with other preliminary costs, the crane pads and laydown space required will be for fewer lifts, due to the larger components used in MMC projects. This is discussed further in Chapter 6 and shows how the cost can be impacted on preliminaries and facilitating works due to time on-site as well as complexity of requirement.

Building Works

Building works encompass the direct construction costs of a project. This category includes the costs of structural and architectural elements such as walls, floors, windows, and doors. In this analysis, particular attention is given to components that could be replaced with alternative construction methods, such as modular production, to assess their cost impact. Building costs are typically broken down into three main components:

- *Labour costs*: Influenced by productivity rates.
- *Material costs*: Dependent on material selection and market fluctuations.
- *Plant and equipment costs*: Costs related to machinery and tools used on-site.

Preliminaries

Preliminaries include costs associated with site accommodation, temporary services, and welfare facilities. These costs are typically time-dependent, meaning that longer construction periods result in higher preliminary costs (Towey, 2018). Modular construction, which offers reduced construction times, can significantly impact preliminary costs. In early project stages, preliminaries are often estimated as a percentage of total project costs due to the limited availability of precise data. And they do reflect the time used on site, as shown previously the reduction in time on site is often calculated when using MMC projects, but the reduction in quantity is not. For example, an MMC project may be 52 weeks long where as a traditional project may be 104 weeks long. This reduction in time will be included when pricing, but the fact that there is a reduced need for site accommodation, offices, welfare facilities, storage facilities, etc. due to the reduction in personnel and materials on site, is rarely considered.

Overheads and Profit

Overheads refer to business expenses rather than direct project costs. These can include administrative expenses, insurance, and operational costs. Since project-specific factors such as the number, size, and duration of projects undertaken within a financial year affect overhead allocation, estimating these costs can be complex. Contractors often apply a percentage-based estimate using historical data.

Profit margins, in contrast, are typically determined by market conditions, as discussed in Chapter 4 higher throughput is possible in a supply chain when using a pipeline system, this can spread overheads over more productivity. Quantity surveyors and contractors monitor industry trends to set competitive profit rates. The adoption of modular construction can influence overhead allocation by reducing project durations. For example, if a company's annual management costs total £200,000 and are spread over ten projects (£20,000 per project), completing 20 projects within the same period using modular methods could reduce the per-project allocation to £10,000, making bids more competitive. However, since modular construction is not yet widely adopted, such adjustments are rarely applied in practice.

Project and Design Fees

Project and design fees include costs associated with architectural design, cost management, structural engineering, and mechanical and electrical (M&E) services. Additional costs may arise from required surveys such as topographical assessments, asbestos inspections, and ecological studies.

While design fees apply to all construction methods, modular production often integrates design and structural elements within the module cost, potentially reducing overall fees. However, this potential cost-saving is often overlooked in traditional cost models. One thing to also note is that the design happens earlier in the project as it needs to collaborate with the manufacturing design team, who need to start production well in advance of what happens on site. It is still probably a cheaper process over all as the manufacturing team can carry part of the design cost (and the more MMC used, such as Modular, the greater the reduction), but the process must start and be frozen earlier. See discussions earlier in this book relating to design freeze (see Chapter 6).

Risk

Risk, in this context, refers to financial uncertainties associated with construction projects rather than health and safety risks. Risk management involves identifying and mitigating potential issues that could impact project costs.

In construction, risk is often accounted for through contingency funds. A higher risk level requires a larger contingency allowance. Modular construction can reduce project risks by minimising on-site work, leading to lower contingency requirements. Similarly, if off-site manufacturers assume responsibility for design accuracy, client-related risks decrease. However, risk reductions are rarely reflected in cost estimates due to a lack of historical data on modular savings.

Inflation

Inflation adjustments in cost estimates account for price changes over the project's duration. Quantity surveyors apply estimated inflation rates based on market conditions. As with overhead costs, reductions in project duration through modular construction are often not reflected in inflationary calculations.

Tax

Taxes in cost estimates are calculated based on applicable rates at the time of preparation. Currently, tax rates for modular and traditional construction methods remain the same. However, tax policies can be adjusted to incentivise modular construction, for example, by reducing Value Added Tax (VAT) on modular components.

By understanding these cost categories, construction professionals can more accurately evaluate the financial implications of different construction methods, particularly modular and panelised systems.

Cost Variable Consideration

As stated above there are various variables to consider, and these are expressed below with an indication of how each may be developed. For ease these are presented in alphabetical order not by order of importance, this is due to the fact that different cost variables will have a different level of importance depending on the site, for example, the impact of weather is likely to be an important factor in the North of Scotland and maybe not as much in the middle of England.

The typical measure of cost impact is a relationship between the amount or frequency of the variable happening, set against the monetised aspect of that variable. Therefore, a variable that happens a lot, and has a high monetised impact is of greater impact, and therefore consideration, than a variable that has a low frequency rate and/or low cost associated.

Accident frequency rate

Accidents, including ill health, is a costly aspect of the construction industry, and with the industry being one of the most dangerous, more dangerous than manufacturing, the change in production method from on-site construction to off-site can have an impact on this cost variable. In its basic terms the accident frequency rate impact is a measure of how often accidents and ill health happen in an environment – the more often it occurs, the greater the impact.

Accidents in the construction sector can happen from a multitude of dangerous activities from working at height, slips trips and falls, working with electrical equipment, especially if the weather is wet at the time, using hazardous materials, etc. The cost of an accident can be very high – while deaths are not as common as in the past they do still happen and this can be seen as the ultimate human cost. On a smaller level of impact, this could still be injuries and ill health that is life changing and the human impact often affects not only the worker or other stakeholder who is injured, but also their extended family and friends. As well as the human cost there is nearly always financial cost associated with accidents and ill health – this comes from the impact of health care, labour replacement, increased insurances, fines, etc.

The frequency of accidents has been shown to reduce through legislation, which often focuses on greater control of working methods or increased supervision. As discussed earlier in this book, one of the key advantages of off-site manufacturing is the higher level of supervision possible within a controlled factory environment. Standardised processes and the repetitive nature of tasks further strengthen this control, contributing to improved health and safety outcomes.

External factors – such as extreme cold, heat, or adverse weather – can increase risks on site (whether through clothing restrictions, ice, or snow) but these are largely eliminated in a factory setting. By moving tasks off-site, projects also reduce the number of exposure hours at height, which in turn lowers insurance costs and lost-time incident costs. This links directly to the quality assurance measures (see Chapter 5) and programme certainty (see Chapter 6).

It should be noted, however, that transferring work to factories does introduce new categories of risk, such as those associated with heavy machinery or production line processes. Nonetheless, these risks remain significantly lower overall compared with traditional on-site construction environments

Air Permeability and Air Quality

Air permeability is the process of air passing through the building fabric in an uncontrolled manner. This can lead to issues with condensation, mould growth, and excessive energy use. Most modern buildings are designed to reduce or eliminate air permeability and relay upon designed in passive and mechanical ventilation systems to control fresh air exchange. However, not all uncontrolled air exchange can be designed out of a building as human error can often play a part in how well the building is designed and built, and therefore there is a tolerance given to the amount of air leakage that is allowed on a building, to pass regulation.

The greater levels of control that can be seen in off-site construction can help with reducing the impact of human error, as well as the increased ability of automation that can be deployed in a factory environment, and should reduce the air leakage. This level of control helps to repeat and enhance air tightness outcomes, reducing the need for re-testing, which also helps with the finances (see Chapters 5 and 6). However, the cost of the air leakage impact is difficult to monetise and is likely to be quite low.

Archaeological Find

Antiquities are a common issue in the construction sector across the world – in some countries there is less of a requirement on the contractor or project to do anything about it, but in other countries, including the UK, finding an archaeological item can prevent the project from moving forward, and can delay project s for months or even years. All of which can have a cost impact.

However, the majority of archaeological finds are discovered in the ground while doing the sub structure work for a project. While there are systems that allow for substructure elements to be manufactured off-site, there is often still the need to excavate and carry out the majority of the work in a traditional way. Therefore, the risk and cost of archaeological find, even though is very high, the change between production methods is minimal or non-existent.

Comfortable Work Environment

The level of comfort experienced on construction sites is improving with technological developments and legal requirements. Comfort, in this instance, is aimed at the experience of the human while working, and this is different to levels of comfort seen at rest. So, the important aspects are to ensure the worker can carry out their duties while staying as safe as possible, but also maintaining an ambient body temperature, maintaining health through sufficient sustenance and water, keeping as dry as possible, and ensuring the environment has the right level so of light and sound in order to maintain the required levels. While on site there are developments in ear protection, lightweight but warm clothing, waterproof clothing, and much improved site office, accommodation so that workers can dry themselves and warm themselves as needed. The work is still carried out in what can sometimes be a harsh environment, and this is getting more so as climate change alters the weather of the world. As such, maintaining the right levels of human comfort is still a difficult task, and this can in itself lead to a slowing in productivity, as the more comfortable a human is the more productive, they can be.

Working in a factory environment it is much easier to maintain the levels of comfort a human needs, control exposure to the weather elements (see below) but also easier to access sustenance and water and safety measures. This will raise the productivity of the production method, but also, when the item gets to site to be installed, and as the installation is so much quicker than on-site installation methods, it is easier to control the human comfort levels when using off-site construction even during the on-site phase.

Construction Cost

This is one of the easier aspects to compare between the productivity methods. The cost of installation of off-site, or on-site items, including the cost of the item itself is take onto account. When looking at construction costs it typically comes down to four key elements, labour costs, material costs, equipment/plant costs and on costs (preliminaries). One of the key aspects in the cost comparison is the on costs, and these tend to be higher for factory-based production than they do for site-based. The on costs will include the cost of the facility – on-site is typically stores and site office and other accommodation, welfare facilities, etc., whereas in off-site the entire factory facility is to be paid for, and this is an expensive undertaking.

There are some benefits, however, such as due to the additional control in a factory environment, and the greater ability to have repetitive work practices, there is more scope to use less skilled workforce, and this can reduce productivity costs. Also, there is greater ability to purchase materials in the sizes needed, cutting down on waste, etc., and the plant and equipment needed can be fixed (such as cranes), and can operate more safely and effectively (due to operating on a smooth concrete floor) and so becomes more efficient. The ability to use plant in places that may not be feasible on site also helps. However, it is assumed that the cost of the production facility in an off-site factory does cause the off-site product to be more expensive that the on-site product when looking at pure construction costs. This appears to be a key reason for the lack of use of the systems in recent years.

One consideration that is explored more below, is the fact that current off-site productivity appears to follow the process of construction, rather than those of manufacturing, and as such, potential cost savings that could happen are not being fully realised currently. This is causing cost issues between product choices. For example, modular productivity is cheaper if standardised sizing is used, but this is not fully understood by designers, as such many modular companies being asked to price for work are having to adjust to non-standard sizes, making them more expensive and even further from competitive.

Construction Energy Usage

Energy is used on a construction project in a multitude of ways, but almost all of them are linked to time. First, energy is needed to power the site accommodation such as offices, welfare facilities, etc. The longer you have the accommodation the more power that is needed. Second, power is needed to produce the project, powering tools, and equipment typically, with the aid of electricity either from mains power or on-site generated power. There is also a safety measure needed from power, for light and heat when working either in the dark or cold. All of which will be reduced if the project time is reduced, for example, using off-site construction.

Another aspect to consider is if the work is carried out using off-site construction. The installation is much quicker and this allows for greater use of conditions that reduce the need for power – for example, with quicker builds, it is possible to plan the work to be carried out on the longer days of summer, taking advantage of more daylight needing less power. Also, the majority of the production methods can be carried out in a factory reducing the need for power tools and equipment on site. A final consideration with regards to power use is the fact that renewable energy sources are easier to utilise in a fixed location – for example, if using a factory to do the production, which is a bit harder on a temporary construction site.

Construction Water Usage

Water is also used for a multitude of purposes on a construction project – it is used for the purposes of maintaining health of the workers on site, but also it is used in the production itself within some of the "wet" trades. These would be things such as plastering, concrete, mortar production, etc. Water is often quoted as the next scarce resource on the planet, and as such a sustainable development will work hard to try and reduce the amount of water used. It is not easy to completely remove the use of water as so many of the products used in the construction industry have properties that are not easy to replicate in other materials. For example, concrete is easy to form when wet, and has considerable strength in compression – when added to a steel frame it also has considerable strength in tension, but it is very difficult to find a material that is not water based with the same properties.

Off-site manufacturing can come in many forms, some of which also use water-based products such as plaster, concrete, etc. but there are also some that use less water-based products such as timber frame, CLT and SIP panels, etc. Therefore, this aspect needs to be considered as part of a sustainable development process. From a cost point of view, sustainability will have a financial impact.

Added to this is the increased need to conserve water as a commodity, and when we look at sustainable development this is becoming more and more important, especially when looking globally. Carbon footprints have been around for a while as the need to conserve fossil fuels has been established – as water becomes more and more of a scarce commodity a "water footprint" measure may soon be common practice. Factory production methods give greater levels of control to ensure all resources are not wasted, if the process is followed correctly.

Cost Certainty

Cost certainty is an aspect of the industry that is considered an important consideration for a client when entering a project, sometimes more important than the cost overall. Cost certainty is the understanding that the price given either at budget stage, or more likely at tender stage, is the eventual price that the client will pay for the project. This is not an easy thing to provide and so often cost certainty is given within an acceptable range. The reason for the difficulty is a combination of things – uncertainty due to ground conditions, and many of the other cost variables listed here – but there is a key impact that is client changes. In the majority of off-site projects there has to be a design change cut off, design freeze or design restriction date, past which clients are not allowed to make any changes to the project design unless major cost is incurred. This is due to the fact that the products are designed and produced in advance of when needed on site, and, due to production line logistics, this can often be well in advance even of the project starting on site. Changes to the design may alter products already built, and due to the production line, the delay could be months and it could cost a lot of money. As such, most clients entering into a project involving off-site will agree to the design freeze conditions, and while this is not the only cause of cost fluctuation it can be a major cause, and so using off-site can lead to projects that provide greater cost certainty. With the production processes being in a controlled environment the impact of some of the other causes of cost fluctuation, such as weather, health and safety, supply chain issues, etc., can also be more easily controlled.

Cost of Re-work

Re-work is one of the most significant yet often overlooked costs in construction projects. At its core, re-work refers to any work that must be completed more than once, generating additional labour, materials, and equipment costs. It also frequently involves dismantling or correcting incorrectly completed work. In some cases, the overall cost of re-work can approach three times the original expenditure, due to the cumulative effect of these factors.

Re-work can occur in several ways. It may arise from incorrect installation or damage during installation, for example, a plumber noticing a chipped toilet pan. Replacing the pan involves sourcing a new unit, removing the damaged item, and reinstalling it. If the defect originated in manufacturing, the supplier may provide a replacement at no cost, but installation-related damage will still incur additional expense.

Another category of re-work arises from latent defects, which only become apparent after installation or even post-occupancy. These defects often incur higher costs because the workforce must return to site, potentially disrupting other projects. Furthermore, rectifying latent defects may require dismantling surrounding work, causing additional damage. For example, replacing a damaged door frame after completion might necessitate repair or replacement of adjacent walls and finishes, a scenario that would be avoided if the defect were addressed during initial installation.

Production methods have a significant influence on re-work. Off-site manufacturing allows components to be produced in controlled factory environments, with higher levels of supervision and quality control, reducing errors prior to delivery. While some human error may still occur, tighter supply chain management and controlled storage minimise the impact compared to traditional on-site production.

A key challenge with re-work is that its costs are largely unrecorded. Subcontractors typically absorb minor errors as part of their operational overheads without passing costs to the main contractor or client. Nevertheless, these costs are real and embedded in project pricing. Reducing re-work through MMC can improve subcontractor profitability, encourage more competitive bidding, and ultimately lower client costs, even if not explicitly documented.

Factory-based production, first-off checks, in-line quality assurance, and integrated sign-off procedures significantly reduce re-work loops and programme knock-on effects. This aligns with the QA gates and templates outlined in Chapters 5, and Chapter 6 – embedding these checks into the programme helps prevent lost time downstream.

Design Change Cost

The design change cost discussed above comes from clients changing their minds and the subsequent impact that has on the production – the later into the production cycle that the change is made the more the impact is likely to be. The design freeze process used in off-site production can help to ensure this does not happen. An important consideration in construction projects is the impact of client-initiated changes after the design freeze. Once a design has been finalised, any subsequent alterations often require input from multiple members of the design team, including architects, structural engineers, quantity surveyors, and mechanical engineers. Beyond these professional fees, changes at this stage can be particularly costly in MMC projects due to the need for re-manufacturing components and managing the associated logistical adjustments. The probability of significant expense increases sharply once production has commenced, as factory schedules, transportation, and site delivery plans must be revised to accommodate the modifications. Consequently, late-stage design changes can have a substantial financial and programme impact, highlighting the importance of timely client decisions and robust freeze-point controls.

Design Cost

Following on from the above, the design costs in a building are typically spent before all, or at least most, of the production starts. The design of buildings can be a costly process, as each building project is unique, as the building has never been built in that situation before. The closest the industry gets to repetition is where multiple storeys of the same building are built, or in housing estates. Even so, the slight differences, location, distance from site office, weather, time of year, etc,. can all cause some slight changes. This number of changes is not common in a factory production.

The costs in design come from paying the professional expertise of the architects, architectural technologists, BIM managers, quantity surveyors, project managers, structural engineers, health and safety coordinators, building services engineers, etc., all of which will cost to access their level of expertise and knowledge. This level is required when each project is unique. If aspects of the project can be standardised, such as with the use of off-site production, then the required need for design input can also reduce accordingly. For example, if every project uses the same bathroom pod as part of its design, then there would be no need for the design cost associated with bathroom completion.

Design Error

Another one of the key aspects that can add costs to a project is design errors – as with design changes, the further into a project the error is discovered, the more costly that error can be.

Errors can come about for a number of reasons – including mis-communication, in that the client or design team did not understand what was being asked for or presented. Also, human error can be a cause – such as members of the design team putting something into the build that should not or could not be there. Mistakes in good faith, such as specifying materials that are no longer available when the build starts, etc., may have a reduced impact if discovered earlier. Using a design process such as BIM can help with reducing the potential for these errors, but also having a design freeze enforced due to off-site manufacturing can also help, as can the requirement for standardising designs to suit the most efficient production method (as discussed above). By using BIM and off-site methods the risk of design error should be significantly reduced, although not totally eliminated.

Design Time

This relates to the direct link between time and cost– the longer it takes to complete the design the more it is going to cost. The more complex the design the longer it is going to take. So, following on from the discussion above, the more standardised processes and products that are used in the design, the less complex it will be and the cheaper the process should be. The use of BIM and off-site should in face lead to a cheaper design cost. In the majority of off-site projects, the design is carried out by the manufacturer to suit the overall project requirements. However, as mentioned above, if standardised sizes can be used by the project designer, this will in turn make the design of the product easier and cheaper overall.

Embodied Carbon

One of the key considerations for a sustainable project is the impact of carbon on the production. One of the key aspects related to this is transportation and logistics – for example, in getting the materials to the site in order for the production to complete. There are several aspects to this that need to be considered, when deciding on production type. Materials tend to fall into two categories: those of primary or secondary (there is also a tertiary category, but that relates to service provision rather than material provision). Primary materials are materials provided in their raw state – for example, some construction sites will now manufacture concrete on site, and have the raw materials of sand, gravel, cement, and water delivered in their raw state in silos. Secondary materials are materials that are manufactured before they arrive to site. Take, for example, a brick – these are manufactured off-site, and delivered to site. Depending on the size of the site, they may be delivered to a local builder's merchant first before going to site.

On a traditional on-site production, all of the materials need to be delivered from the location they are either extracted as a raw material or manufactured. If they are manufactured, then the carbon footprint also has to consider the delivery of the raw materials to the manufacturer. As many construction sites are reasonably temporary and have a multitude of different materials, the carbon footprint for all of these materials can be quite high. One way to reduce this is to have much more of the products pre-manufactured before coming to site, as would be the case with off-site construction. Sensible manufacturing plants would be very close to the raw materials, cutting down on carbon footprint as well as the cost of transportation of the materials they need to manufacture. With fixed locations this is easier to accomplish. For example, it would make sense for a brick manufacturer to site their factory production facility close to a clay pit. If off-site manufacturers can do the same, this will help with reducing the carbon footprint.

There is, however, one further consideration for the carbon footprint and often stated as a reason not to use modular products – the fact that in modular the transportation moves large volumes of air, as the product is manufactured in three-dimensional complete units, all of the air inside the unit is transported as well as the materials. Whereas with many of traditional materials the materials are stacked without air in between, consider a pack of timber, or plasterboard. Panelised systems made off-site can combat this as the panels are often stacked flat with minimal air gaps in between making them much more carbon efficient. Carbon footprint can help to drive specification changes, and this could impact on cost of products that needs to be considered.

End of Life Recyclability/Future Adaptability

This is a recent consideration in construction projects, often called the circular economy. The basic premise of the circular economy is that when designing a project, and in particular when specifying materials, it is incumbent on the design team to consider what will happen with each of the items in the build when the building comes to end of life. Quite often a building will be demolished or at least considerably refurbished while many of the products are still of a standard that can be used for many years to come. As an example, a house may lose its aesthetic appeal and so needs to be demolished, but the brickwork, slates on the roof, some of the timbers, etc., may still be in good enough condition to be repurposed on a different building.

With off-site construction there is a new dimension emerging, with the idea that modular products can be added and removed as the building needs changes. Again, take a house – perhaps when the house is bought by a young couple starting out in their life together a two-bedroom bungalow may suit their purposes, but as they start a family, they may want to expand their house upwards to make room for the growing family. As they get older and their children leave home, they may want to remove these again back to the two-bedroom bungalow, and perhaps add on extra rooms to allow their aging parents to live with them. This flexibility is seen as a good by-product of using modular build in standardised sizes.

Energy Efficiency

As well as the discussion on carbon footprint above, the build of a project will also impact on the amount of energy used to run the building. The lower the thermal transmittance of the building fabric, the greater the levels of efficiency that can be achieved. By improving the building fabric U values (thermal transmittance), less heat energy either leaves the property in cold weather, or enters the property in hot weather – this will reduce the need to either mechanically heat (using heating systems) or cool (air conditioning) the building. There are other aspects that can be considered in reducing energy use, such as using renewable energy systems, but the decision to use modular or on-site production will not impact on the decision to utilise renewable energy sources. The simulated virtual environment that is possible from using BIM will help to ensure maximised potential of energy efficiency.

Therefore, the consideration between the two systems is if one can achieve greater levels of efficiency than the other. Traditional brick and block buildings would normally use a wide cavity and fill that cavity with insulation, whereas a panelised system uses the gap between the wall uprights to insulate – this can have one or two effects, either the greater level of insulation can be achieved, or a grater footprint inside the building can be achieved. For example, using 100mm of insulation in a cavity would need the cavity to be 125mm wide which would reduce the internal footprint of the building (due to having thicker walls) whereas a timber frame system with insulation built in the wall would not need to have this much wider cavity, and so the footprint can be increased. Another consideration is the greater levels of accuracy in the installation that can be achieved in a factory production facility as opposed to on site, and slight gaps where insulation does not cover will reduce the impact of the insulation, and these are more likely to occur in the less controlled on-site production. In other words, the repeatability enabled via factory production and off-site techniques will reduce thermal bridging errors, improve energy efficiency and life cycle costs, particularly energy bills, should fall. Another consideration is that panelised systems and modular systems can more easily be designed to allow for reducing thermal bridges, this is a point where a less efficient material bridges from the inside of the building to the outside of the building. For example, a timber stud, a block wall, brick tie and brick external wall, are all thermal bridges. But by staggering the vertical members in a panelised or modular system, this can be reduced.

Estimating Error

As with design error above, the human element used in estimating can also lead to errors, in a similar way – misunderstanding systems, discussions, information, etc., can lead to mistakes. Estimating errors are typically dealt with via contractual arrangements but can still lead to a cost impact. As with design error it can also lead to a reduction in work relationships on the project as well, which can also impact on cost. With estimating errors, it is either by the consultant Quantity Surveyor during the cost planning phase of the project, or by the contractor during the pricing phase.

If the error is showing the cost to be too high, then the impact could be that the project is stopped, or the design is altered unnecessarily. It may even lead design team to move from one production type to another due to the error, for example, estimating the cost of modular production in correctly and so deciding to use on-site methods instead. The lack of knowledge and understanding of systems will have a big impact on estimating errors, and this is one of the key issues with off-site, as discussed earlier in this book. The example given off by just altering the size of the modules slightly having a big impact on the module costs is relevant here.

If the error is too low similar decisions can be made in error and causing the same problem. With decision making, it is key to get as much information as possible, within the limitations of the decision, so that as close to the correct decision can be made. Limitations typically include time to get the information, or lack of knowledge, but unless this is done in the right way, then errors will be made, and these can be costly.

Typically, an error made earlier in the project (during the design stage) can be more easily rectified if found early but can have the biggest impact. Errors by the design team typically require the client to pay. However, errors in the tendering process may be harder to rectify but will typically be less likely to impact on the client as contractually it is often the contractor's responsibility to cover mistakes they have made. If the mistake is big enough to cause a catastrophic failure of the company involved, however, then there will be a financial impact on the client as they will need to find a new contractor to complete the work, and this usually comes at an increased cost.

To sum up, errors of any kind are likely to impact on the client and project cost at some point.

Financing Cost

Financing is a critical component of any construction project, with the goal of generating a return on investment (ROI). This return can be financial, as in private or speculative developments, or non-financial, such as the social value seen in public infrastructure projects. Regardless of the type of return, the method of financing significantly impacts project risk, cost, and profitability (Myers, 2017).

There are three main ways to finance a construction project: self-funding, external loans, or hybrid approaches. Each method carries varying levels of risk, costs, and benefits (Higham, Bridge, & Farrell, 2016).

Self-Funding

Self-funding involves the client or developer using their own capital, such as cash reserves or liquid assets, to finance the project. While this method avoids interest payments, it carries opportunity costs. These are the potential gains forfeited from not investing those funds elsewhere, such as in interest-bearing accounts or shares. Additionally, using internal funds can weaken a company's short-term liquidity by transferring liquid assets into less accessible, long-term assets. This shift may increase financial vulnerability if unforeseen costs arise.

Another drawback of self-funding is the loss of potential interest income. In high-inflation environments, such as the UK in 2023, savings accounts offered returns as high as 6%. Choosing to fund a project rather than save at those rates can represent a significant financial sacrifice.

Bank Loans

Bank loans are a widely used method for funding construction projects, but they typically come with relatively high interest rates. In addition to interest, the opportunity cost must also be considered, this refers to the potential financial benefits lost by not using the funds elsewhere, such as in savings or investments. When combined, interest charges and opportunity costs significantly increase the overall cost of borrowing.

Over the course of a project, interest is usually calculated on a compound basis. This means that as spending increases throughout the build period, interest accumulates not only on the original loan amount but also on the growing balance. As a result, the longer a project takes to complete, the more costly the borrowed funds become.

While some loans require repayments to begin immediately, others may have repayment terms that start after the project is finished. Delaying repayments can further increase the total cost, especially if the loan continues to accrue interest during this period. The total funding cost can therefore grow substantially over time if not carefully managed.

This highlights the importance of time in project financing. Longer build times and deferred repayments lead to higher financing costs. Understanding and planning for these effects is essential in managing project budgets and ensuring financial sustainability.

Reduced Sale Price Funding

Reduced sale price funding is another method used by developers to raise money before a construction project begins. This is often done by selling some of the homes "off-plan", meaning they are sold before they are built. To attract early buyers, developers offer these properties at a discounted price.

The size of this discount can vary depending on the company, the type of property, and the local housing market. In popular areas where properties sell quickly, or where the developer has a strong reputation, only a small discount may be needed. In less in-demand areas, or where buyers are more cautious, a larger discount might be required to secure early sales.

Although this method provides early income that can help fund the start of the project, it comes with a downside: the discount represents a permanent reduction in the potential profit. Unlike loans, where repayments might be covered by future sales, the money lost through discounts is not recovered.

Developers must carefully plan how many properties to pre-sell. Selling too many at a discount reduces the overall profit of the development. On the other hand, selling too few means the developer may need to rely more heavily on loans or internal funds, which can increase risk or cost. If too many properties are sold early, it can also send a negative signal to potential buyers, suggesting the developer may not be financially stable, which might then require even bigger discounts to encourage sales.

Because the income from off-plan sales is received before construction begins, it is treated differently in financial planning. The cost is considered a fixed reduction in revenue, rather than an ongoing or future cost like loan interest. As a result, it is included in overall project financial models when calculating total income and profit over time, often using tools like net present value.

Net Present Value

The time value of money is a key financial concept in construction. Future income is less valuable than immediate income due to inflation and the potential earning capacity of money. Therefore, project financing must also account for net present value (NPV), especially in developments with long-term returns like rental properties (Myers, 2017).

Income Generation

The income for a construction project can come in several forms: through property sales, rental agreements, or through use by the final client (the end user). Each of these types of income is affected by time, so when the method of construction changes the build time, it can also affect how and when income is received.

In many real-life developments, income is considered separately from costs. However, for a full understanding of project funding and decision-making, it is important to link them together and explore how income timing affects overall financial performance.

Sale Income

One of the most common sources of income is through selling the completed development. This can happen in a few different ways:

- *Speculative sales*: The developer builds the project without having a buyer already in place. This approach carries more risk, but it gives the developer full control over the design, which can lead to higher profit if the market is strong.
- *Pre-arranged sales*: In this case, a buyer is already lined up before the project is finished. This reduces the risk for the developer and often means the money is received more quickly after completion. However, there is still some risk if the buyer withdraws from the deal.
- *Off-plan sales*: Here, a buyer agrees to purchase the property before it is built, often at a discounted price. This brings in money early, but at the cost of reduced profit.

A key benefit of sales income is that it usually comes in a large lump sum once the property is sold. This means the developer can pay back loans or other costs quickly, especially when compared to rental income.

Rental Income

Rental income usually comes from properties that are meant to be rented out over time—like social housing, office spaces, or private rental homes. These developments may or may not have a tenant lined up before construction is complete.

One main difference with rental income is that it tends to start later and builds gradually. Instead of a large payment all at once, rental income comes in smaller, regular amounts. Over time, this can result in more income than a sale, but it takes longer to recover the money used to build the development.

Client as End User

Sometimes, the client who commissions the building is also the one who will use it. In these cases, income doesn't come from selling or renting the building, but from the activities that happen inside it.

For example, a hotel will make money through bookings, while a factory might produce goods to sell. This kind of income is harder to measure in terms of how it pays back the development cost, since it depends on many other factors. In most cases, the increase in overall profit for the client is used to estimate the value of the development.

Other Considerations

Some developments don't have financial income as their main goal. These are often projects that bring a social benefit, such as new schools, hospitals, roads, or public transport systems. While these might not generate income directly, they improve the community and can lead to broader economic benefits.

To understand the full value of such projects, a Social Impact Assessment (SIA) is often used. This looks at things like how the project improves access, its flexibility for future use, its impact on health, safety, comfort, and energy use, and how it affects the local area.

Finally, it's important to note that switching from traditional to modular construction methods does not always increase income for the end user. While modular methods might shorten build times or reduce environmental impact, most of these benefits are seen by the developer or contractor, not the buyer. As a result, the use of modular construction usually does not lead to a higher sale price or rent.

Time has a big impact on how much a construction project costs, especially when it comes to funding. If a project looks at how money going out (costs), money coming in (income), and time all connect in the same financial plan, it can help with better decision-making. This kind of planning helps show how delays or changes in schedule can affect the project's budget.

However, in many real-world cases, especially in fast-moving or competitive markets, developers may not include this time-cost link in their calculations. Because of this, the real impact of time on costs can be hidden or ignored.

While the type of funding used (like loans or personal investment) also affects costs and risk, one of the biggest ways to reduce overall costs is to shorten how long the project takes to build.

A shorter build time usually means less interest to pay on loans, better cash flow, and a quicker return on the money invested. Therefore, the production selection method can be key when exploring the impact of choosing the correct funding method.

Waste and Hazardous Waste Generation

Waste on a construction site can cause damage to the environment, can be a health and safety hazard, and can cause significant cost – therefore finding a way to deal with waste effectively is really important. Waste comes in many forms and from many sources, and while this makes the management of waste complicated, it also offers opportunities to make the project more sustainable in lots of ways. Typically, these can be broken down into waste disposal and waste production.

First of all, disposing waste in the right way is important, so a multilayered skip system is important so that waste materials can be appropriately stored and then disposed of. As much waste as possible should be recycled, reused, repurposed and there are many ways to do this. Unused materials can be sold back to the merchant, or it could be moved to the next project and used there, it could be given to local community projects or local schools and colleges, or it can be sold off in a form of a yard sale process. Most of these allow the reuse of the materials and saves money from putting materials into landfill sites, and reduces the cost associated with skips, disposal fees, and taxes

Another consideration is ways to reduce the production of waste in the first place, and one of the many benefits to using BIM as a system is to also reduce waste production. If it is possible to know exact measurements of each item, and those items are standardised, then the waste production can be eliminated. Take, for example, plasterboard that comes with the standard size of 2.40m high board while the most common UK ceiling height is 2.3m. Thereby every sheet of plasterboard produces some waste. Again, standardising sizes as can be done effectively with off-site construction techniques, will help to eliminate waste production.

Finally, a consideration is the waste that is produced in the factory production area, as there seems little point in moving the waste production from on-site to off-site, unless there is a further benefit. In a factory setting, where standardised sizes are being produced more regularly and in a fixed location, it is easier to try and remove waste at source, so standard lengths of timber can be ordered from the manufacturer, and this means less waste comes into the facility in the first place. Standardized practices and improved control will also prevent much of the accidental waste (this is where mistakes are made during production causing an item to become waste). As the facility is fixed it can also be easier to dispose of waste in a more efficient way, such as it being easier to create links with local communities, or some materials (such as timber waste) can be used in other ways on a fixed location, such as being used of heating purposes, etc.

Health and Wellbeing

Health, Safety and Welfare (H, S & W) is a critical and well-studied topic within the construction industry. It affects not only individual workers but also companies, clients, and society at large. Costs associated with H, S & W can be divided into preventative and restorative categories.

Preventative Costs

These are planned and included in the project budget, typically under preliminaries in the New Rules of Measurement 2 (RICS, 2021). Examples include:

- Personal protective equipment (PPE)
- Scaffolding and harnesses
- Welfare facilities

These costs are known, time-related, and decrease with shorter project durations. As such, they are outside the scope of restorative cost analysis.

Restorative Costs

These are costs resulting from accidents or ill health on site, and include:

- *Lost time*: Workers may need time off, leading to a direct halt in productivity. Delays can affect the critical path, extending project timelines and damaging reputation.
- *Sick pay*: Most construction workers are self-employed and do not receive sick pay. Those under PAYE may receive statutory sick pay, but this is limited.
- *Loss of production*: Absences may reduce efficiency. Replacement workers may be unfamiliar with the site, slowing progress. Injured workers returning may experience mental strain and reduced performance (presenteeism).
- *Overtime and temporary labour*: Bringing in temporary staff or paying overtime increases costs and may result in lower productivity due to fatigue or unfamiliarity.
- *Reputational damage*: Accidents can lead to negative media and public perception, especially through social media. This may result in fewer future work opportunities and legal scrutiny from bodies like the HSE.
- *Project delays and LADs*: Delays can trigger Liquidated and Ascertained Damages (LADs), which are penalties written into contracts (e.g., 0.5% of the contract value per delayed week).
- *Investigation costs*: Investigating an incident takes time and involves several staff, disrupting productivity.
- *Fines and legal costs*: Although not calculated in this context, employers may face significant fines if found in breach of health and safety laws.

Other Considerations

- *Material and equipment damage*: Accidents may damage buildings, tools, or machinery, creating additional delays and repair costs.
- *Emotional and social costs*: Accidents can have a lasting impact on individuals, families, and communities, including increased reliance on local services and changes to lifestyle.
- *Industry structure*: Construction work is less repetitive and more mobile than manufacturing, leading to higher accident risks. A transient, self-employed workforce has less consistency in training and awareness, increasing risk.

Incomplete Survey Detail

This cost variable links to the discussion earlier on design errors, if the information gathered at the start of the process is inaccurate then this will lead to errors in the design with all of the consequences listed above. Some of the information that is gathered and that can cause issues include topographical surveys – if these are inaccurate then the assessment for quantity of ground removal will not be correct, and foundation design may be wrong. Also soil surveys assessing the make of the ground can cause similar issues. Flora and fauna surveys carried out can be affected by the time of year and some of the preventative and relocation practices needed, if this is incorrect, can lead to major natural environment and ecology issues. This impact would happen whatever the production system used, and, in most cases, it causes problems while groundwork is undertaken and therefore is less likely to differentiate between production types. However, one thing to note is that due to the early need for procurement on an off-site project, any errors have less time to be discovered before incurring higher impact.

Investment in Local Community

As stated above, investment in local communities is a big part of what the construction industry does – when considering a sustainable development, community is one of the three key pillars considered (along with economy and environment). Part of the process of investing in the local community is increasing jobs, etc., for local construction companies, and an increase in trade for local businesses as the building is built. When using off-site systems, the positive impact on the local community can be reduced as the time that the workforce is on site is reduced, and the number of smaller sub-contracting firms needed is also reduced. In the factory location, however, the opposite is true – however, what this does is benefit the local community of the fixed factory by a great deal, whereas the mobile workforce nature of on-site production is likely to have a small impact but in multiple communities.

Discussed below is the impact of pollution, and this is also a consideration for local communities, as construction sites can produce a lot of local pollution.

Job Security

Linked to the above is the concept of job security – this is becoming less of an issue in modern times as mobility of labour is more common. Mobility of labour is where people are more likely to move job for reasons of either changing or staying in a set geographic location or moving into a new job sector. A few years ago, it was more common to stay in the same job and location for life, and so job security was more of a consideration.

With the transient nature of construction projects with shortish timespans for projects, the ability to provide longevity of work on a project is limited. However, with the increase in use of a sub contract supply chain this is becoming less of an issue, as the smaller companies only need to find work in their immediate area, and this can be on multiple projects.

The fixed location of modular is better at providing a more long-term job security solution, as the factory is fixed, although, as stated above, this can have an impact on the sub-contractor companies who rely on the on-site production.

Added to all of this is the consideration that we are currently in a skills shortage situation, and this is reducing the potential impact discussed here. Off-site production with the reduced need for higher skilled workforce is making it easier to complete construction projects in areas where the subcontractors needed are not readily available. Also, the lack of available skills also drives up the cost of construction which will make it easier to justify a cost comparison between off-site and on-site.

Lack of Brief Development

The brief is arguably one of the most important aspects of a construction project's success – it is the key document produced in the early stages of the project that sets out what the client needs. All subsequent decisions are based on this initial document. It is to be noted that the document can still change as the project progresses, but the key principle is to control decision making. If the brief is not sufficiently developed at the time of needing to order the off-site components this can lead to errors as discussed in the design error section above, so it is even more important to get this brief established in detail early, in projects of this type.

Lack of Planning Approval

One of the key trigger points in a construction point is whether planning permissions are given or not. Planning permission is a process where the secretary of state approves local development plans for each of the county and city councils, which will then provide a basis for a set number of each type of development. For example, County Council A may require 1,000 new homes, 500,000M^2 of retail space, and 1,000,000M^2 of commercial manufacturing space, etc. This is then used by the councils to approve or decline planning requests – therefore if the client asks for permission to build a housing estate of 30 homes, it is more likely to get approval if it falls within the development plan requirements. The next level down is then more about aesthetics, where the drawings, elevations and visuals for a project are submitted, and must meet the local requirements – typically this is in relation to the surrounding area. Then, finally, there are the subsequent requirements such as impact on highways, the environment, local resources, etc. – for example, a new development that will place too much of a burden on the local school, NHS, draining systems, etc., will be declined; if access to the development off the local highway system is dangerous it will be declined.

As such the production type is not always a big consideration with planning permission, although in the regard for increased traffic, in getting, in some cases, much bigger deliveries to the site, it may become an issue.

Less Manual Labour

One of the key considerations, and alluded to in other parts of the book, is the ability to automate production more easily when in a fixed location, with a stable base to work from, as can be seen in a factory production setting. Automated processes – such as using robotics and laser cutters, etc. – will significantly increase the level of accuracy and quality control in the manufacturing setting. The more a product is bought, the greater the companies' ability to invest in automation, and therefore the greater the benefit to the product.

Many errors, whether by accident or lack of care, can happen due to human input – the more this can be reduced the greater the quality is likely to be.

Localised Pollution Issues

Localised pollution typically falls into one of three main categories:

- *Dust* is probably the easiest to recognise, as construction activities often create lots of dust as the work includes the use of materials such as sand, cement, gypsum, etc. that are powdery by nature. When working with these it can cause dust to enter the air and settle in the surrounding areas. If not controlled this can cause issues with the health, wellbeing and lifestyle of local communities. There are a number of ways to try and control dust on site, including wheel washes, solid hoarding, damping down of activities, etc.
- *Noise* is another form of local pollution that needs to be considered, this is often reliant upon the location where the work is taking place, and the hours of the day that the work is carried out. For example, work carried out between 9 and 5 on weekdays in a rural location is not likely to have a major impact, whereas a 24 hour a day project next to a residential area will have major implications. The ability to prevent sound in activities is quite difficult and so good stakeholder interaction is also key on construction projects
- *Light* is the final key local pollution, while there are many more these three cause the most disturbance locally. The light comes from vehicles and artificial light sources, as well as some reflection and glare, both from the construction process and the finished building. Again, there are things that can be done to reduce the impact such as blocking of the windows, so internal light sources do not get out, or maximising the day light working option by doing more summer work and les winter work. But again, it is really complex to prevent all light pollution and so good stakeholder engagement is also key.

Off-site construction has the potential for a big positive impact on the above local pollutants, as the work is mainly carried out away from site and brought not site, the delivery and installation, can be planned to only happen at sensible times. Also, the speed at which the production can go up using off-site, means the impact locally will be more short term, and this will have a positive impact on local and community nuisance.

Longevity/Durability

As discussed above when discussing the circular economy, the longevity of building materials is a key consideration but often does not align with the longevity of the building itself. One of the key considerations with regards to any new or innovative technology is the lack of detail surrounding the working life of a material. And while off-site construction is not a new system, and many of the products are still in a working state from many years ago, it is considered a new technology by the industry. Often the uncertainty over the longevity of the items used is a key source of risk for developers, contractor and clients when exploring off-site as a production option.

Programming and Time Considerations

The processes for creating and monitoring construction programmes in traditional projects is well established – for example, work break down structures and precedence diagrams to ascertain the activities that are to be planned is followed by a network diagram to help plan the critical path activities, and the Gantt chart to help in easier monitoring of the project when underway.

The programme is used to ensure that resources are delivered in time to carry out the work and meet the expected completion date – the resources include labour, plant and materials. When looking at the differences between traditional and off-site there are a few considerations.

First, traditional construction processes are generally made up on many different sub-contractors and suppliers for many different materials. When using off-site construction this can be reduced, and when using modular it can be reduced to only a few, as the majority of the build comes from one source. This has many benefits in that trying to organise the delivery and insertion from one source is easier to organise. Although it is to be acknowledged that the supplier has a big supply chain and this could impact on their ability to deliver to schedule, with it being in a fixed location and producing the same sort of items on a regular basis, this is less risky than on a traditional construction site.

Second, and linked to the above, off-site manufacturing is not well established in the UK, especially when you look at the more complete versions seen in pods and modular. As such, if there is an issue with the delivery then there is little chance of getting the product elsewhere in time to maintain the programme on the main build. In traditional construction, the materials and supplies used are mainly standard materials that can be obtained from a number of different merchants, therefore a late delivery or a failure of one of the suppliers can usually be remedied quickly and with little financial impact. If the company delivering modular systems goes into liquidation, then getting a replacement can take many months due to the production line of the limited number of suppliers. Again, this will potentially ease, as the systems become more widely used and more manufacturers join the supply chain.

Finally, the production line needs to be considered when ordering – again this is more important the more complex the off-site system, as production can take a while, and as modules will be bespoke to the particular programme needs (while still maintaining the standardised size and approach). As such, the need to order early is important and is linked to the design freeze requirement mentioned above (see Chapter 6 for more detail).

Poor Quality Management

This aspects links to the entire project, including design, pre contract and post contract activities. Quality control in all processes is a complex issue, but much of it comes from correct supervision and checks. To maintain the quality of the design and the design team, they meet regularly and meet with the client to ensure the design is meeting the needs of the client. Poor design decisions can lead to the need to redo the design, or if discovered too late on the project it can also cause a delay in the programme and additional costs to the client (as well as poor reputation for the design team). The design carried out by the off-site production team will be improved due to the standardised nature of the products – however, there is a risk of complication between the project design team and the manufacturing design team. If the project design team are not fully aware of the processes of off-site construction, there is a chance of low-quality design and design error between the off-site product and the rest of the project.

Later in the project there is a need to monitor quality during the production phase which will include the monitoring of the materials when they arrive to site, and the activity of the installer when putting the material in place. With the use of off-site construction panelised systems this is also the way they are checked, but with more complex modular systems it is considered good practice to inspect the quality of the product prior to them leaving the factory as if there are any issues it is easier, and cheaper, to fix them at source. They will then also be inspected upon arrival on site and following installation. These additional checks will be more likely to ensure off-site products are of the correct quality when installed. And the additional and enhanced quality checks that are enabled by producing the work in a factory environment should ensure greater quality control overall.

Overall, it is assumed that off-site products will provide a higher quality than on-site, but there is a need for greater levels of knowledge of the systems, by both the design team and the on-site project team to ensure this high-quality level is met. Consideration can be given to the adoption of the 5.7A controls and 5.8 routines from Chapter 5 to shift re-work from cost to prevention.

Preliminaries

Preliminaries on a construction project are the elements of support that a project needs to run, but in the main do not stay on the project when it is completed. It will include items such as site offices, welfare facilities, temporary services, temporary roads, security hoarding, etc. Preliminaries are mainly priced on a time basis – the longer the project, the more the preliminaries will cost. For example, a site accommodation will be priced based on the number of weeks it is n site.

The cost associated with preliminaries is usually well understood by contractors when tendering for a project, as it is closely linked to programme duration. As noted earlier, the time required to complete a project is generally predictable, and therefore preliminary allowances can be adjusted with relative confidence. The complexity arises when applying this to different production methods, particularly where knowledge of MMC processes remains limited. As familiarity increases, more accurate adjustments can be made.

At present, many off-site projects simply reduce preliminaries in line with the overall programme savings. However, the reality is more nuanced. Not only is there a reduction in time-related preliminaries, but also in the overall quantity required – for instance, fewer welfare facilities or traffic management provisions may be needed due to shorter site durations. Importantly, realising these savings depends on aligning site operations with factory production cycles. Effective use of 4D planning and integrated scheduling tools allows site rhythms to be matched to factory batch deliveries, ensuring that programme compression translates into tangible cost reductions. As explored in Chapter 6, such integration enhances certainty, enabling preliminaries to be reduced in both duration and scope.

For example, site accommodation and welfare facilities on a traditional project will provide facility for all of the sub-contractors that will be on site at one time. In an off-site project, particularly a modular project, the number of sub-contractors is reduced. Storage is also less complex, and the need for temporary services is reduced. As soon as this increase in knowledge is used, then the cost of preliminaries will reduce accordingly.

Productivity

Productivity in the construction industry is lower in the UK than many other industries and has other variables feeding into it. It relates to the variance in productivity when working on-site compared to off-site. Some of the aspects that may impact on this could be, for example, weather related – it is not as easy to work when the weather is wet or cold or too hot, whereas production in a factory environment is more controlled. There is also an aspect of reduced productivity when looking at control – for example, in a factory environment the management and control or monitoring of activities is direct and close thereby improving productivity, whereas on site, the production activities can be quite a distance from the control source (site manager) and may be irregular thereby reducing productivity. In a factory environment, due to the standardised practices that are available, more use of plant and machinery can be utilised increasing overall productivity. On a site this would not be possible (it is rare to have cranes on low level domestic sites) and so most of these activities are carried out by hand and are less efficient and can cause health and safety issues, poor quality and damage to materials.

The key difference when using off-site, though, is the speed – manual labour is quicker installing large items than smaller items. For example, it takes almost as much time to lay a brick as it does a block, and the block is six times the size of the brick. If you can expand this principle and install an entire wall section instead, then productivity will increase.

Programme Certainty

Following on from the discussion on programming earlier, this concept is to do with the accuracy of the programme. One of the issues mentioned above is the number of different components, suppliers, and sub-contractors that are involved on a construction project. When you reduce this with the use of off-site construction, especially when using modular, then this complexity is reduced and therefore the accuracy of the assumptions also increases (see Chapter 6).

Off-site construction, however, is a production method that is not fully understood by site managers and project managers, and so getting the accuracy right on a project that is not known is difficult. An added complexity in using off-site, when it comes to programme accuracy, is the reliance on one supplier for large parts of the build – this puts the control of timings on a project into someone else's control. If something happens during the production of the modules or panels, it is the responsibility of the supplier and manufacturer to fix it and meet the agreed deadlines. The ability to go to another supplier and get the item, if the delivery schedule slips, is not an option with such large items from such a small supply chain, as was discussed earlier.

Regional Economic Uplift

Construction projects are diverse and unique and due to their nature are built in the location that they are required. As they are so diverse, one of the key impacts of having a construction project locally is to increase regional income. As contractors and sub-contractors gain from doing the work, the employees working for them get paid and typically spend that money locally, in retail and other living expenses. On major projects this impact is quite high and can last for a number of years and over a much wider geographical area, as bigger projects will draw contractors from further afield than smaller projects.

When using off-site, this is not as impactful, as the majority of the work can be carried out in a fixed location away from the actual site itself, and the majority of the sub-contractors who would normally benefit locally do not. As such, the impact for an off-site project, depending on the level of off-site used, can be felt in the surrounding area of the factory rather than the locality of the site itself. Also, as will be discussed more below, a higher level of skill is required for working on-site than in factory conditions – this is due to the more controlled and standardised working conditions available in a factory environment. Higher wages also tend to go alongside higher skill levels, making the impact more noticeable. Manufacturing plants should be located close to a port to reduce the embedded energy and carbon of its raw materials, and the major ports of the UK have lower labour costs. Therefore, the social benefit of having a major construction development in a location can be reduced as a lot of the labour is moved to the factory location, but with careful consideration, governments can focus the benefit in the areas that need it most.

Repeatability and standardisation

This concept has been alluded to a few times already in the book, but the basic concept is that for the major benefit of off-site to be realised, including the most financially beneficial, then standardised sizes should be used. For example, with panelised systems production facilities, saws, jigs, etc., are all set up to build to a standard height. If a different height is needed all of these need to be reset and this adds to the cost and the potential for error. If project design teams are aware of this when designing, then they can design to meet the standard sizes of the manufacturing facility, keeping costs and complexity down. When using modular processes this is more impactful as three-dimensional sizes are used and if these are standardised then it can be more beneficial. By adding 5mm on to the width of a module, or height of a module it can take the cost from being competitive to uncompetitive.

As these sizes become more standard in the sector, then the manufacturing facilities will be able to work on more standardised and repeatable processes, making them more accurate and cheaper (see also Chapters 4 and 5). This is seen within the industry at the moment with standard sizes of doors and windows being incorporated into designs, thereby keeping costs down. It is possible to go to window and door manufacturers and get non-standard sizes which will cost more, but it is rarely done. The concept, if transferred to other items such as walls, floors, roofs, and entire section of a building, will produce cost benefits.

This standardisation will also make the potential for re-use and the circular economy much easier – as the design sector uses standardised approaches, then modules and panels can be removed from buildings and reused on other buildings.

Re-work Impact

One of the key aspects to using modular systems, as opposed to traditional, is often cited as an increase in quality. One of the costs associated with diminished quality is the need to return to a project and repair or replace the poor-quality item, known as re-work and can be between 5 and 20% of the project cost. Poor quality has been reported as being prevalent in the construction projects, and in particular the residential sector. As re-work carried out there is not only the impact of replacing the materials and having someone to carry out the work for a second or in some cases third time, but also identification of the fault (snagging), management/supervision and control of the re-work or repair to cost, and a reduction in reputation for each time re-work is carried out. An often-overlooked aspect of re-work is when subcontractors are not called back to site, but notice the work is sub-standard while doing the work. For example, while installing an item the worker may notice it is damaged, or damage it during installation, requiring a re-installation – this will slow productivity as it is replaced. Quite often costs associated with returning to site to complete re-work is not recorded anywhere and appears as a hidden cost.

The use of off-site should reduce the re-work impact, although it is still unlikely to remove it in its entirety. See the section on quality management above for more discussion on why this may be the case.

Risk and Risk Management

Monitoring, assessing, and controlling risk on a project will help to ensure project success as a high-risk project is not an attractive environment for anyone to work in, despite the potential for bigger returns. The less risky a project is, the more comfortable contractors are in providing lower profit margins – if the contractor knows they will get 3% nett profit they are more likely to bid for the 3%, whereas if it is not certain if they will get 0% or less, then they are more likely to bid for 5 or 6% instead. Therefore, removing the risk on a project is to everyone's benefit. The process for risk management is similar, no matter the level of risk. First of all, assess the risks, and grade/rate them, then apply some form of control measure and re-grade them. Finally, it is important to apply the control measure and monitor its impact to learn for future projects. With the repeatability of off-site, this learning is quicker and more beneficial.

Health and safety, as discussed above, is one of the key risks on a construction project and using off-site should reduce this risk, although not remove it entirely. However, there are other risks on a construction project outside of the risk to health and well-being. For example, the risk the client may go into liquidation, or the contractors, or the supply chain, there are risks of unforeseen events in the ground, or the risk of increases in material costs, etc. – the list of risks is quite extensive, but in the majority of cases, the shorter the time spent on site the less that risk is, and the quicker the project can proceed the shorter the exposure of the stakeholders. Therefore, it can be seen that in many cases the use of off-site can create a less risky environment. It is to be noted however, that having one major supplier so heavily relied on, and with such a small supply chain, if something happens to that one supplier the impact can be very big. Often this level of risk can put contractors off using the systems available. Contracts should recognise factory slots, QA holds, logistics as time-risk items (see Chapters 3 and 6 for more detail).

Scalability

As the number of off-site units produced annually grows, then the rules of economy of scale (Myers 2017) would suggest the cost of these will reduce. It is, however, a sliding scale – as production increases a factory unit will eventually reach capacity and have to open a new production line. This, in turn, will reduce the financial impact of economy of scale as the new production line is not at capacity, and so on. Economy of scale has been discussed above, but it is important to also consider that the ability of organisations to scale up on productivity will be easier with factory production than it is with on-site production. Part of this is due to the additional control and standardised production processes that can be established in a factory setting, but also the reduction in the need for higher labour skills means there is a bigger employment pool to use when the need to upscale occurs.

There is a need to consider production facility when upscaling, although to begin with factory production can move to a 24 hour a day production using shift patterns, which can triple the capacity for a factory. This is something that is not always easy to do with an on-site process, due to local pollution issues as discussed above. Often factories are in areas that are designed for factories, such as business parks and logistic centres, etc., and so extending a factory or moving to a bigger facility in the same location is often reasonably easy to do.

It is assumed in the sector that as systems become more viable, then the economy of scale with make them even more viable, leading to growth in that sector. This can be seen with the impact on standardised systems already used such as timber frame systems, and trussed roof manufacturing. It is also to be noted, that once transportation issues are developed the ability to move into new geographical areas becomes easier for manufacturing companies, as delivering a module 20 miles away is the same process to delivery it 100 miles away. As such, this makes the exploration of new markets easier to do, including exporting to other countries. This is not the same when using on-site systems, as they are so labour intensive – new geographical markets typically means gaining specific knowledge of the local supply chain.

Skills Gap

The world of construction is currently suffering from a skills shortage, as fewer people see the industry as a viable career path – this is a process that is just getting worse. Off-site construction is seen as a solution to this. There is also a known gender, cultural, and ethnic background gap within the construction industry, at both point of entry and during promotion processes. While this gap is reducing and the industry itself has responded to ensuring the need for a more inclusive environment, it is still assumed by the influencers of certain cultures and genders that it is not the best place to work. The same issue does not appear to be as prevalent in the manufacturing sector – as such, off-site construction may be attracting a more diverse workforce than on-site.

Also as discussed several times above, due to the standardised production processes that can be created in a more controlled environment, a less highly skilled workforce is needed to carry out the production process than are needed on a construction site. There is still a need for higher-level skills in the design, cost, and project management processes. As such the skills gap is also not as noticeable within the manufacturing sector than it is in the construction sector

Social Licence

Social licence is the acceptance of an organisation to operate and has many influences. A lot of what will help with social licence is related to local investment as discussed above, as a fixed factory that helps to improve the local economy off-site manufacturing plants are typically accepted. The on-site project, however, can have a different issue with social licence, and some of this will depend upon the type of project that is under development. If the local community are happy for the project to be built, then the social licence will improve – the unstated issue is that quite often it is the contractor and developer who suffer from social licence, rather than the client that commissions the project. As an example of this with house building projects, it is often the developer that suffers the bad press, although as discussed above it is the local council that allows the development, and it is not the developer who ends up using the facility. The developers, especially on speculative developments, would not build unless there is a local need, as if they did, they may not sell their end product. Often it is not the level of need that influences the social licence, it is more of an issue with having the development close to areas of society that do not want it. As a "not in my back yard" (NIMBY) situation, for example, there is an agreed need for more schools, and prisons, and leisure developments, new housing estates, etc. – however, most communities do not want to have these facilities built in their immediate area.

Other aspects that will influence social licence are the speed of development – in the UK there is a public perception that if something is built quickly it is of lower quality. With modern production methods this may not be true but there is still this perception. Also, if developments do not work within sustainable development protocols, then there is a risk that governments will force them to do so via legal changes, etc.

This concept of social licence means that more consideration of local population and communities is needed to ensure the project runs as smoothly as possible. While off-site can lead to faster build processes, the fact that there is less activity on site, less local pollution, etc., should lead to better engagement with local communities, and better social licence. There will be a need to mitigate the impact of transportation as discussed above, and there is a need for strong community engagement, as with most construction projects, to ensure the concept of off-site is fully understood and accepted. Generally, social licence should be easier with off-site production methods.

Soil Investigation Mistakes

The ground conditions, as discussed above, are a key aspect of many construction projects and they are also a complex aspect to investigate. The ground conditions will determine the type of foundation that is required to hold the building, and if they are too weak it may stop the project altogether, as the cost of stronger foundations, such as pile foundations, may make the overall project cost too prohibitive.

The typical process used to ascertain the make of the ground is to either use bore holes or trial pits. This is an excavation into sections on the project site, to see what type of soil there is. However, it is not possible or viable to dig up the entire site, and so these tests are carried out in certain strategic areas. Sometimes, the sporadic nature of these tests does not expose the areas of weakness that may occur in the soil, thereby only discovering the weakness of soil once the excavation begins. If this weakness is insurmountable, then costs will have been occurred to start a project that cannot be completed.

As this work is carried out in the ground, it often will not impact on the differences between on- and off site production, except perhaps for the consideration of the weight of the off-site structure compared to the on-site equivalent. However, one thing to consider is that in off-site projects much more expense is committed to earlier in the project than with off-site – if the project is halted then this is a greater loss than may have been experienced with on-site techniques.

A further aspect to the investigation int eh soil is the risk of UXO (unexploded ordinance) such as bombs, etc. left over from World War 2. However, the risk is similar to that as discussed above, and unless off-site sub-structure components are used, there is limited potential impact other than the risk if the project fails to move forward.

Supply Chain Partnerships and Resourcing

The complexities of the supply chain have been addressed in more detail in other parts of this book, however, from a cost point of view, it is important to consider the key differences of the two production types. With on-site there are many more sub-contractors working on a project, all in close proximity to each other, leading to logistical complexity and sometimes increasing the risks. When using off-site, there are usually fewer sub-contractors which makes the complexity less but also reduces the competition which can lead to higher overall costs. When there is a small supply chain pool to operate in, should something happen to the sub-contractor/supplier being used on a project, this impact is much bigger than would be seen on an on-site project. Therefore, much stronger financial checks need to be put in place when there is a smaller supply chain.

One of the key benefits that can be seen with off-site production is the ability to develop stronger relationships within the supply chain, enabling volume pricing and stable lead times (see Chapter 4). The manufacturing facility is stationary, and so suppliers in that locality will be used regularly, whereas the mobile nature of construction means that when using on-site techniques the suppliers may be different for each project. With a smaller supply chain, the clients, developers, and contractors can develop stronger links with the manufacturer. This partnering ethos can lead to better knowledge of each other's work practices, leading to greater savings and more efficient and beneficial working environments. For example, see the discussion above on standardising sizes of development to meet the standard off-site sizes. All of this can have a positive financial impact.

Tendering Cost

Tendering is the process by which clients appoint contractors, or by which developers and contractors appoint sub-contractors. In a traditional development, there are numerous trades and material streams that must be procured, coordinated, and managed. Many trades are locally engaged for the short duration of a project, while materials are typically sourced through national suppliers, who in turn draw on international production. This extended chain introduces multiple layers of cost, complexity, and risk.

By contrast, modular systems operate with a smaller, more consolidated supply chain. Typically, this is limited to groundwork, external cladding, landscaping, and the module itself. The modular manufacturer, being factory-based (unless using mobile or "flying" factories), is able to exercise greater selectivity in procurement, and in some cases can purchase directly from international producers depending on scale. This offers the potential to reduce both costs and complexity.

On a positive note, these advantages can be reinforced by ensuring clarity in Employer's Requirements (ERs). Including MMC-specific details, such as QA evidence, factory production calendars, and logistics constraints, helps reduce uncertainty for tenderers, thereby lowering risk allowances and improving value outcomes.

The addition of partnering processes as discussed above, will also reduce tendering costs further, as organisations get used to working together, the speed and accuracy of pricing mechanisms also improves. One further consideration is that the lack of competition in the supply chain currently can lead to closer project tenders overall as competing contractors will often have to use the same subcontractor/supplier and are often given the same cost price to work with. If the off-site system being used covers the majority of the project build, say, for example, with modular, or with off-site bridges, tunnels, etc., then the client will have less need to analyse financial differences between tenders and can focus more on quality differences.

Vehicle Movements

One of the key health and safety risks on site is the number of vehicles required and moving around the site at ethe same time, at the same time as the workforce. This risk requires complex and costly traffic management systems to ensure productivity is kept as high as possible while the workforce is kept safe. With using off-site construction, especially the more complete modular systems, this traffic is reduced as often items are lifted into place using a crane system. As most materials are installed in the factory there is then limited need for on-site traffic to get the materials to the place of installation.

With vehicles, it is also worth noting the consideration of getting materials delivered to site, that has been discussed in more detail in the embodied carbon section above. The consolidation of logistics can help reduce aspects such as weather impact or downtime, and it is important to take this into account when devising delivery buffers.

Weather

Weather has an impact on the efficiency and cost effectiveness of construction activities, whether they are carried out externally or internally. The impact of weather on UK industry is growing in severity, and is expected to get worse over the next 30 years. Weather is split into 13 categories consisting of clouds, snow, thunder and lightning, frost and ice, rain, fog, humidity, hail, hurricane, wind, tornado, storms, and temperature. Productivity is generally impacted by fatigue of the worker, worker impeded (for example, by excessive clothing), activity impeded (for example, by having to protect the product from the weather), and emotional impact (the worker is slower or faster depending upon their emotional state) – as such the fact that the sky is covered with clouds will not have a detrimental effect of any significance on any of these productivity factors.

When discussing the impact of weather it must be considered alongside the current expectation of a rise in temperature of the Earth and the effect this will have on the climate for the future. While the research is based on current weather patterns and probabilities, an increase in climate temperature (between 1.2°C and 4°C is currently predicted in the next 30 years), will increase the chance of extreme weather occurrences, and not just extreme heat. The years 2022 and 2025 had several heatwaves where the temperature was over 30°C for several days at a time and a drought was felt by many, but the rise in temperatures is as likely to bring with it excessive rain and extreme cold days. With some predicting more droughts and flash flooding in future years, this will increase the impact to consider here.

With the longer on-site production there is more chance of weather impacting the work. Weather comes in many forms, and the impact can be both physical and psychological when exploring the impact on the productivity of labour. There are further impacts on the use of a variety of materials as some, such as concrete and mortar, will not cure correctly if installed in certain low temperatures, some paints will not adhere properly if there is excessive humidity, etc. There is also an impact on some plant/equipment, such as strong wind impacting on the ability of cranes to operate, or some mobile machinery operating in heavy rain. Finally, there are risks associated with weather which can prevent some work when the risk is increased beyond a safe point – for example, frost and ice when working at height, or high temperatures when installing metal panelling.

Costing Processes

The above are all of the cost variables that will need to be considered during the costing process in relation to both on-site and off-site production methods and these have given an indication of how each will impact on the overall cost. However, knowing this is just the first part – the next aspect is to attempt to monetise these cost variables, and, to date, this is where industry and academia have struggled.

The costing process most typically used in the construction industry is a variation on activity-based costing, and it links the resources being used for each of these activities. The key resources used are labour, plant, and materials, and the process is to ascertain how much of each is used on any given activity, multiplied by the cost of each and that will give an indication of the cost. As already discussed, there is more complexity to it than this (otherwise we would not need quantity surveyors in the industry) – such as adding in waste for materials, or converting time-based cost for labour and plant – into element-based cost that is required for activity-based pricing. As an example, the activity-based cost unit for brickwork is M^2, in other words how much does a M^2 area of brickwork cost? The materials resource is already costed using these unit rates (it is possible to work out how many bricks are needed in a M^2), but labour and plant is costed based on time, and so a knowledge and experience-based conversion is needed to alter the cost to match the M^2, to show how long it takes for a bricklayer to lay a M^2 of brick, etc.

This is the standard way to calculate costs for a construction project, and it does work with off-site – however, due to the unknown cost aspects listed earlier in this chapter, and due to the lack of knowledge of estimators and contractors in the use of off-site, it is rare that the true and accurate cost is taken into account when pricing off-site projects. As the knowledge of off-site and its impact grows then the knowledge of the cost aspects will grow, and this knowledge will increase the levels of accuracy with the costing. However, with current practices, off-site project costing is less likely to offer it as a viable solution. That is due to the fact that the above variables are rarely taken into account. Work by Roberts (2025) on the specific costs of weather impact has shown a potential saving of between 2 and 5% just from this one variable. Re-work has been shown to have a saving of between 5 and 20% on project costs. The reduction in design change aspects, the reduction in risk, etc. all have a potential for a financial impact.

The modern estimator, quantity surveyor, commercial manager, and contractor need to be aware that the costs associated with the use of off-site are more complex than at first appears. While processes are developed to provide a more accurate assessment of cost impact for each of these processes, then it is important to at least take them into consideration.

The process for calculating the impact is complex in some areas, hence the reason it has not been done clearly to date, but in its basic format it looks at the probability of something happening and the financial impact of it happening. So, for example, if we can say the cost impact for re-work to a toilet cistern is roughly 50% of the cost, and there is a 25% probability of that re-work happening, then the cost impact can be calculated at 12.5% of the project cost (this is an example as to date no one has worked out the probability or costs associated with re-work).

Below is an example of weather impact. Roberts (2025) analysed five years of weather data over the United Kingdom, for the five main weather types (light rain, heavy rain, excessive high temperatures (over 25°) and excessive low temperatures (0°) and wind (over 15mph gusts). As it was deemed these were levels that can impact on the work carried out on site. The next step in the research explored what the productivity impact is from each of the weather types. The issue with this is that it is subjective, each worker will be impacted in a different way by exposure to different weather types. In cold weather, some go slower as they feel emotionally drained and find the work more cumbersome in lots of additional clothing and protective gear, whereas others may worker faster in order to get the work done and get out of the cold.

On top of this complication there is not a standard rate of pay that is paid to workers in the industry, and so to get a standard impact of weather on a construction project is complicated. This has lead Roberts (2025) to develop a formula that can be adopted by contractors to assess the impact of weather on their workforce. This in turn will add to the narrative around off-site or on-site production techniques.

Simplified Calculation Method

The aim of the calculation is to estimate how weather impacts productivity on a project, based on when activities fall within different seasons of the programme. The process can be broken into two main stages:

1 Work out how the project duration overlaps with the seasons

- First, establish the start month of the project, entered by the contractor.
- Then, split the total duration (d) into three equal parts (thirds).

So:

- Months m_1 … mk = first third
- Months mk+1 … m2k = second third
- Months m2k+1 … md = final third

Example: For a 15-month project starting in January (d = 15, k = 5):

- First third = m_1 to m_5 (Jan–May)
- Second third = m_6 to m_{10} (Jun–Oct)
- Third third = m_{11} to m_{15} (Nov–Mar).

Each season (winter, spring, summer, autumn) covers three months. To measure overlap, we use:

- S_{ij} = This is the season $_j$ value of month $_i$. If this = 1 then the month is relevant in this calculation, and it would equal 0 if it is not relevant.
- j has four possibilities (winter, spring, summer or autumn)
- i has 12 possibilities (Jan, Feb, Mar, Apr, May, Jun, Jul, Aug, Sep, Oct, Nov, Dec)
- The season count has two aspects to it, that is = C (3,4) that is there are three thirds to the build and 4 seasons as possible options.
- C_{pq} – season count for the $_{pth}$ third (outcome can be 1st, 2nd or 3rd) for the season q (outcome can be winter, spring, summer or autumn)

The season count is then calculated for each third:
This tells us what fraction of each season falls into each third of the project.
Worked example:

- In the Jan–May first third (5 months):
  - Winter = Jan + Feb = 2/3 → 0.66
  - Spring = Mar + Apr + May = 3/3 → 1.00

So, for the first third the project has 0.66 Winter + 1 Spring.
The equation for this looks like:

$$C_{1q} = 0.33\sum_{i=m_1}^{m_k} S_{ij}$$

The second third equation

$$C_{2q} = 0.33\sum_{i=m_{k+1}}^{m_{2k}} S_{ij}$$

And the final third

$$C_{3q} = 0.33\sum_{i=m_{2k+1}}^{m_{3k}} S_{ij}$$

2 Apply seasonal probabilities to activity impacts

Now we estimate how much this seasonal exposure reduces productivity for each activity. The formula is:

$$C_{t1} = \sum_{k=1}^{5} W_{ij} P_{ak}$$

Where:

- t = the project third
- W = probability of a weather type in a given season/location
- i = location
- j = season
- P = percentage impact of that weather type on a given activity
- a = activity (e.g. masonry, roofing).

Summary of the method

1 Split project duration into thirds.
2 Count seasonal overlaps for each third using *C1q, C2q, C3q.*
3 Apply weather probabilities × activity sensitivities to each season.
4 Sum the results to get an overall productivity adjustment per activity.

This makes it possible to model how different start dates, durations, and locations change productivity due to seasonal weather.

The equation for the final part looks like this

$$C_{t1} = \sum_{k=1}^{5} W_{ij} P_{ak}$$

Chapter Summary

This chapter has explored the cost impacts of modern methods of construction (MMC), with a focus on how off-site approaches reshape both direct and indirect project expenditure. A central theme is that costs are not eliminated but shifted: MMC moves expenditure from the uncertain, fragmented environment of the construction site to earlier and more controllable stages in the factory and supply chain.

Re-work illustrates this shift clearly. On-site, mistakes often go unrecorded, absorbed into subcontractor overheads or hidden in programme slippage. In contrast, factory-based production benefits from first-off checks, in-line quality assurance, and integrated sign-off procedures that reduce re-work loops and avoid knock-on delays. Similarly, safety risks are diminished when tasks are moved off-site, lowering both insurance premiums and lost-time costs. Though new types of risk exist in factory settings, these remain far less disruptive than their on-site equivalents.

Financing is another area of adjustment. Faster completions enable earlier income generation, but this advantage is coupled with the requirement for earlier capital drawdown, shaping client decisions on cashflow and net present value. For planning authorities, programme compression offers reduced traffic management costs and road possession fees, but it also introduces the need to balance accelerated delivery with local development targets. Communities benefit from reduced disruption through fewer deliveries and shorter site durations, though public perception of rapid delivery must still be managed through engagement and education.

From the contractor's perspective, supply chains consolidate under modular construction, reducing complexity but placing new emphasis on Employer's Requirements. Clarity on QA evidence, factory calendars, and logistics constraints helps to limit risk allowances in tendering. Preliminaries, meanwhile, cannot simply be reduced proportionally by time saved – true savings arise when factory production cycles and site rhythms are integrated through 4D planning and coordinated calendars.

In summary, MMC shifts cost from late, uncertain site conditions to earlier, more controllable stages. The gains arrive through predictable timeframes (Chapter 6), reliable supply partnerships (Chapter 4), and disciplined quality controls (Chapter 5). Where these systems are present, cost outcomes improve and become repeatable. Over the coming years, it is hoped that more academics, researchers, and industry will extend the evidence base for these cost variables. Until then, responsibility rests with decision makers to adjust production and installation costs realistically, accounting for the full spectrum of benefits that MMC delivers.

References

Higham, A., Bridge, C., and Farrell, P. (2016). *Project Finance for Construction*. London: Routledge.

Myers, D. (2017). *Construction Economics: A New Approach*. New York: Routledge.

NAO (2007). *Homebuilding: Measuring Construction Performance*. London: National Audit Office.

ONS (2024). Construction statistics annual tables. www.ons.gov.uk/businessindustryandtrade/constructionindustry/datasets/constructionstatisticsannualtables (accessed 5 October 2025).

RICS (2021). New rules of measurement. www.rics.org/profession-standards/rics-standards-and-guidance/sector-standards/construction-standards/nrm (accessed 7 October 2025).

Roberts, D. (2025). *Critical Analysis of the Impact of Weather on the Cost related to Productivity of Traditional and Modular Construction Techniques*, unpublished PhD thesis. Aston University.

Towey, D. (2018). *Construction Quantity Surveying: A Practical Guide for the Contractor's QS*, 2nd ed. New York: John Wiley and Sons.

Van Vuuren, T. J. and Middleton, C. (2020). *Methodology for Quantifying the Benefits of Offsite Construction*. London: CIRIA.

Chapter 8

Concluding Thoughts

Introduction

This chapter serves to draw together the key themes and issues explored throughout the book, with the key intention of turning lessons learnt int action. At the start of this chapter, Table 8.1 shows how these key lessons discussed can be used to enhance the impact of MMC on future construction projects. The detail, evidence, and case studies presented previously are essential to understanding the conclusions reached here, and readers are encouraged to revisit those chapters to gain the full context.

At the same time, this concluding chapter has value in its own right. It brings together several of the core ideas in a holistic manner, highlighting the interconnections between procurement, modern methods of construction, and wider industry outcomes. By clarifying and reinforcing the debates presented earlier, it offers students a structured overview of the sector and a clearer sense of how the different strands of discussion fit together. In doing so, the chapter aims to provide not only a summary, but also a foundation for critical reflection and further study.

This book has explored a range of construction methods available to the industry, recognising that in practice projects rarely rely on a single approach. More often, a hybrid mix of methods is required to achieve the desired balance of time, cost, quality, and flexibility. Each method offers its own strengths, but also presents particular limitations and risks. Understanding these trade-offs is essential for both practitioners and students of construction management.

Stakeholder Considerations

Below is a table summary that will provide key considerations for each of the stakeholders when working with or wanting to work with MMC projects, this is discussed in detail below, but this table can be used as a quick reference.

Modern Methods or Traditional

Traditional construction methods remain dominant across much of the UK industry, but they present a number of well-recognised challenges. These methods are heavily reliant upon a site-based workforce, drawing on a large pool of skilled trades and specialist contractors. This, in turn, requires a complex supply chain and careful on-site coordination to bring together multiple materials, systems, and people at the right time. While this approach has been refined over many decades, its dependence on human factors and logistical complexity creates significant risks and inefficiencies.

DOI: 10.1201/9781003570431-8

Table 8.1 Key Stakeholder Activity

Stakeholder	*Key Consideration 1*	*Key Consideration 2*	*Key Consideration 3*
Client	**Early Engagement and Design Finality:** MMC demands decisions and coordination much earlier than traditional methods.	**Cost, Programme, and Risk Management:** Understand the balance between higher upfront costs, quicker delivery, and supply chain risks.	**Quality, Warranty, and Lifecycle Value:** Ensure MMC solutions meet quality expectations, regulatory compliance, and long-term asset performance.
Main Contractor	**Design and Procurement Integration:** Collaborate early with designers and manufacturers to align design and procurement strategies	**Programme and Logistics Planning:** Carefully manage site readiness, just-in-time delivery, and interface tolerances.	**Quality, Safety, and Compliance:** Coordinate quality control between factory and site, and ensure health, safety, and regulatory compliance.
Sub Contractors	**Understanding MMC Systems and Interfaces:** Be aware of how your trade connects with modular or prefabricated elements.	**Installation Precision and Timing:** Work within tighter tolerances and installation windows, with minimal scope for onsite changes.	**Training, Tools, and Communication:** Ensure operatives are skilled, equipped, and in sync with the main contractor and design intent.
Architects/Designers	**Design for Manufacture and Assembly (DfMA):** Adapt design processes to suit modular or prefabricated systems.	**Coordination and Early Decisions:** Finalise layouts, details, and interfaces early to align with factory processes.	**Aesthetic vs. System Constraints:** Balance design vision with the limitations and efficiencies of MMC systems.
Structural Engineers	**Structural and Service Integration:** Ensure systems and load paths align with MMC components and site works.	**Tolerance and Interface Design:** Design with precision to accommodate tight tolerances.	**Collaboration and Compliance:** Coordinate across disciplines early to ensure buildability and regulatory adherence.
Quantity Surveyors/ Project Managers	**Procurement and Cost Planning:** Due to the changing nature of how and when costs are occurred plans and advice must be adapted.	**Quantification and Valuation:** Mechanisms needed to take into account in factory checks, links to production processes and shorter production times.	**Risk Management and Contract:** Plan for new risks, such as single point of failure, transport and storage issues.

One of the most frequently reported issues with traditional construction is late completion. Projects are highly vulnerable to delays because of their dependence on a large site workforce, where productivity can be affected by illness, fatigue, stress, or adverse weather. Alongside this, the reliance on extended supply chains increases the likelihood of late deliveries or shortages of key materials. These delays are costly, not only in financial terms but also in reputational damage for contractors and clients alike, and they can disrupt wider development programmes.

A second key issue is low and inconsistent quality. Rework, the process of correcting or replacing work that has been completed incorrectly, is commonplace within traditional construction projects. Errors may arise from human factors such as fatigue or miscommunication, but also from material-related problems such as damage during transportation or poor site storage conditions. While some errors can be corrected during the construction process, many are not discovered until much later, sometimes only after handover to the client. These *latent defects* are particularly damaging, as they require teams to revisit completed projects, disrupting ongoing work and adding considerable financial cost. Furthermore, quality problems undermine client confidence and highlight the limitations of site-based processes, where so many variables are difficult to control.

This is not to suggest that traditional construction is without merit. It offers flexibility, particularly for bespoke or small-scale projects, and it draws on a wide range of existing skills that are deeply embedded in the industry. However, the recurring issues of delay, quality shortfalls, and inefficiency provide the context for the increasing attention being given to modern methods of construction (MMC). In contrast to traditional approaches, MMC aims to reduce reliance on site-based labour, simplify supply chains, and create more controlled conditions for production – themes that are explored further in subsequent sections.

In response to the recurring challenges associated with traditional construction, Modern Methods of Construction (MMC) have been increasingly promoted as a potential solution. Advocates of MMC argue that shifting key processes into a factory-controlled environment can help to resolve problems of quality, time, and efficiency. Within factory conditions, production is less exposed to variables such as weather and site-related risks, enabling greater consistency in output. This, in turn, provides stronger quality assurance, as work can be subject to rigorous inspection and testing before arriving on site. Alongside this, the fact that much of the work is completed off-site reduces the overall programme length, as site preparation and module fabrication can often be carried out in parallel. The result is the potential for shorter project durations, reduced disruption on site, and more predictable outcomes.

However, it is important to stress that MMC is not a universal solution in itself. As has been highlighted throughout this book, the benefits of MMC depend heavily on how it is planned, procured, and integrated into the wider project. Without careful coordination, early design freeze, and alignment of supply chains, the advantages can be significantly reduced, and in some cases, MMC may even introduce new risks or inefficiencies. For example, if modules are ordered too late, or if designs are altered after manufacture has begun, costly delays and rework can result. This underlines the point that MMC is not simply a "plug-in" replacement for traditional methods, but a different way of delivering construction that requires corresponding changes in strategy and management.

Another key point is that MMC is not a single technique, but a spectrum of approaches. It ranges from panelised systems and structural insulated panels (SIPs) through to volumetric modular units, light-gauge steel framing (LGSF), and hybrid solutions. Each approach carries its own strengths and limitations, and not every method is appropriate for every project. For example, high-rise residential developments may benefit from volumetric modular solutions, while low-rise housing schemes might be better served by panelised timber systems. Choosing the right form of MMC is therefore crucial to achieving the desired outcomes.

For the industry to genuinely improve performance, delivering projects that meet client expectations while also reducing costs, improving quality, and shortening delivery times, a balanced approach is often required. This may involve combining traditional methods with selected MMC techniques, tailored to the project's specific needs. Equally, success depends on raising awareness and understanding among decision-makers across the sector. By increasing knowledge of MMC's opportunities and limitations, and by embedding this understanding into procurement, design, and delivery practices, the industry can begin to move forward in a way that is both innovative and sustainable.

Off-site structural components, such as pre-cast concrete beams and columns, follow similar principles, with speed and quality being the main advantages. Equally, off-site non-structural components, ranging from cladding panels to service pods, bring comparable benefits and limitations, reducing site labour requirements but introducing dependencies on manufacturing schedules and logistics.

Table 8.2 Summary of Key Decision Characteristics

Traditional construction	is characterised by a highly skilled, labour-intensive workforce operating on site. One of its key advantages lies in its flexibility: it allows for bespoke design solutions and provides scope for changes during construction. This makes traditional techniques particularly attractive for complex or unique projects where adaptation is necessary. However, these strengths are offset by significant drawbacks. Traditional construction is generally slower, highly reliant on large supply chains, and prone to human error, which can lead to rework, delays, and higher costs.
Modular construction	involves the factory-based manufacture of three-dimensional units, often entire rooms, which are then transported to site and assembled, typically by crane. These modules can be produced to varying stages of finish, from structural shells to fully fitted units. The principal benefits of modular methods are speed of delivery, enhanced quality through factory-controlled processes, and the potential for reduced project costs due to programme savings. Yet there are also considerable risks. Rework is especially expensive once modules have been produced, while disruptions in the supply chain, such as the failure of a manufacturer, can create costly delays given the relatively small pool of suppliers. Logistics and delivery scheduling are also critical, as even minor coordination failures can affect project timelines.
Panelised systems	offer a more flexible alternative. Here, two-dimensional components such as walls, floors, or roof panels are manufactured off site and then assembled on site. This approach accelerates construction compared with traditional methods, while still requiring finishing work once the panels are installed. Costs are broadly comparable with traditional builds, as higher manufacturing expenses are offset by quicker on-site installation. Panelised systems share many of the risks seen in modular construction, but the supply chain for panelised products is broader and replacement of components is easier, making this option more attractive to a risk-averse industry.

Emerging technologies are also beginning to reshape construction practice. 3D printing, though still experimental, holds significant promise. By automating the production of building elements, it reduces reliance on manual labour and offers high levels of precision. Although not yet mainstream, the rapid pace of development suggests that students and professionals should remain alert to its future role in the industry.

Other innovations fall under the banner of labour-saving methods. These include systems designed to increase productivity on site through larger or more efficient components, such as oversized masonry blocks, brick-slip panels, or plastic tile substitutes. While these approaches improve speed, they often involve higher upfront costs. Similarly, innovative labour-reducing technologies, such as robotics and automation, are being developed to replace or supplement human labour on site. Their chief advantage lies in increased productivity and consistency, but the substantial capital investment required may act as a barrier to widespread adoption in the near term.

Taken together, these construction methods represent a spectrum of possibilities, from traditional approaches grounded in human skill to cutting-edge technological innovations. The challenge for industry practitioners lies not only in selecting the most appropriate method for each project, but also in combining approaches in ways that maximise efficiency, quality, and value for money while managing associated risks.

Non-negotiables

When choosing methods of production for a construction project, there are a number of important factors that must be carefully evaluated. This knowledge is crucial not only for the design and project teams but also for the client, as effective decision-making should be based on the specific demands of the project rather than simply repeating what worked previously. By understanding these considerations, stakeholders are better placed to achieve the potential benefits of Modern Methods of Construction (MMC), particularly in relation to improvements in time, cost, and quality outcomes.

One of the most significant influences is the requirement for an early design freeze. Because many MMC processes involve off-site manufacture, design decisions must be made and fixed earlier than in traditional construction. This means that any late design alterations can be disruptive, costly, and in some cases, impractical once factory production has begun. Recognising this early-stage commitment helps to frame the wider discussion of how procurement, contracts, and tendering practices must adapt to support MMC.

The procurement and tendering route is another central issue. As discussed in Chapter 3, a range of options exists, such as Design-Bid-Build, Design and Build, and more collaborative models, and while none is inherently "better" for MMC, the timing and integration of decisions are critical. For example, if off-site methods are considered only after design is largely complete, the benefits of standardisation and manufacturing efficiency may be lost. Similarly, in design and build projects, unless the contractor and supply chain are engaged early with clear MMC knowledge, the opportunities for efficiency and quality improvement may be reduced.

Closely linked to this is the importance of contractor selection. The level of knowledge and experience a contractor has with MMC processes will directly influence project outcomes. Without sufficient expertise, projects risk delays, increased costs, or failure to achieve expected quality levels. Selecting partners with proven MMC capability increases confidence that risks will be well managed, and benefits properly realised.

Alongside procurement and contractor selection, contracts and legal frameworks must be carefully considered. Standard forms of contract can often be adapted, but MMC projects frequently demand clauses that address issues such as design lock-in, factory-based quality assurance, and logistics planning. In addition, new approaches to warranties and assurances are emerging, with accreditation schemes and product guarantees giving clients and financiers greater certainty about performance and durability. These tools are becoming increasingly important in building confidence across the industry and will be explored further later in this chapter.

At this stage, it is clear that effective decision-making for MMC depends on more than simply choosing between traditional or modern approaches. It requires an appreciation of how procurement strategies, contractual mechanisms, and early design choices interact, and how these can be aligned to support successful delivery. The following sections will build on these themes by examining the specific influences, such as design freeze requirements, contractual obligations, and emerging warranty models, that practitioners must be aware of when planning to use MMC.

Supply Chain Management

Following the establishment of procurement processes and the appointment of stakeholders, the effective management of the supply chain becomes a crucial consideration. As discussed in Chapter 4, construction supply chains are complex and layered, ranging from the procurement of relatively small items (such as fixings, nails, and screws) to the engagement of major subcontractors responsible for significant elements of the build. They also extend beyond material supply, incorporating design services, cost management, structural calculations, surveys, and even rework related to latent defects once a project is complete.

In traditional construction, the breadth of the supply chain provides a degree of resilience; if one supplier fails, alternatives are often available. For example, should a kitchen manufacturer cease trading during a project, another supplier can typically be sourced with limited delay. However, in the case of specialist MMC solutions, such as modular construction or emerging technologies like 3D printing, the supply chain is often narrow and less established. If a modular manufacturer were to go into liquidation mid-project, developers could struggle to find a replacement, potentially delaying the scheme to the point of making it no longer viable.

This heightened vulnerability means that supply chain selection for MMC requires additional diligence. Financial resilience and operational stability of suppliers must be carefully assessed, as any disruption could have significant implications for time, cost, and overall project viability. While financial checks are important across all forms of construction, they become particularly critical when the supply base is small and highly specialised. Consequently, supply chain management in MMC is not simply about coordinating logistics but also about safeguarding against systemic risks that are less pronounced in traditional approaches.

Quality Management

One of the most frequently cited benefits of MMC is its potential to deliver higher levels of quality compared with traditional construction. Yet, as shown in Chapter 5, this benefit is not guaranteed; it depends heavily on the programme, the systems of quality control in place, and the extent to which these systems extend beyond factory production to include transport, storage, and installation.

There are several reasons why factory-based environments offer advantages for quality management. First, supervision and oversight can be more effective. On-site supervisors are responsible for large, dispersed work areas, and as projects rise above ground level, direct monitoring becomes increasingly difficult. By contrast, in a factory environment, supervisors and quality managers can oversee production from fixed, centralised locations, often supported by technology such as strategically placed CCTV systems.

Second, standardised production processes are easier to achieve in controlled environments. On construction sites, variability in materials, site conditions, and even weather can undermine consistency, despite the use of RAMS (Risk Assessment Method Statements). In factories, standardisation can be more rigorously applied, with materials cut and assembled to uniform specifications and with greater opportunities for automation. This reduces the likelihood of human error and enhances repeatability across multiple components.

Third, environmental conditions are much more stable in factory settings. Site-based construction is exposed to weather, daylight restrictions, and temperature variations, all of which can reduce accuracy and productivity. For example, achieving perfectly plumb brickwork is far more challenging in heavy rain or snow. By contrast, factories can control lighting, temperature, and air quality to create optimal working conditions, not only supporting accuracy but also extending productive working hours. In addition, renewable technologies such as solar power can be integrated more efficiently into fixed manufacturing facilities, supporting sustainable production without the limitations of temporary site-based systems.

Finally, labour conditions differ significantly. On-site construction often demands long hours and overtime to meet deadlines, leading to fatigue, stress, and a higher risk of errors. In contrast, factory-based production is better suited to shift working, which enables more consistent working patterns and can contribute to improved worker wellbeing alongside sustained output.

However, these benefits should not be overstated. The improvements in quality associated with MMC can be quickly undermined if robust quality assurance systems are not extended throughout the entire delivery process. Components may be manufactured to a high standard in the factory, but if they are damaged during transportation, stored incorrectly, or installed without sufficient care, the overall quality of the completed building will still be compromised. Effective quality management, therefore, must be holistic, covering not only production but also logistics, site installation, and aftercare.

Time Management

The principles surrounding time management in MMC projects can be understood through two key points.

The first relates to time savings on site. When off-site construction techniques are employed, certain aspects of the work are completed before components even arrive on site. The greater the level of completeness achieved prior to delivery, the more time savings can be realised. For example, a panelised system may reduce some programme time, but a fully volumetric modular approach, particularly where modules are delivered with high levels of internal completion, can shorten project durations significantly. These time savings, however, must be balanced against the higher upfront cost of more advanced prefabrication, as explored further below.

The second principle concerns the earlier start to design and production activities in MMC compared to traditional construction. In conventional projects, mobilisation typically occurs only a few weeks before on-site work begins, with specialist items occasionally requiring earlier procurement. In contrast, MMC demands that major orders are placed well in advance and that a design freeze is implemented to avoid later alterations. While design changes in traditional projects may cause delays, these are often manageable. By comparison, changes made once MMC production is underway can be highly disruptive, leading to significant additional costs and programme overruns.

As highlighted in Chapter 6, a number of strategies can be adopted to manage these risks, but the key point is that MMC shifts the timing and sequencing of project activities, requiring early commitment and coordination to achieve the intended benefits.

Cost Management

Chapter 7 explored cost in depth, focusing on the variables that influence whether MMC can offer financial advantages over traditional construction. Importantly, it did not conclude with a simple answer of whether MMC is "cheaper" – rather, it highlighted the complexity of comparing the two approaches.

Traditional cost models are built around resource and activity-based analysis, combining labour, plant, and material costs with allowances for overheads and profit. When compared in isolation, on-site production often appears less expensive than factory-based alternatives because it avoids factory overheads. For example, the cost of building a wall on site may initially seem lower than using pre-manufactured timber panels.

However, such comparisons often fail to capture the wider cost variables that significantly influence overall project expenditure. Some of these are partially considered but not fully quantified. For instance, preliminary costs, such as site accommodation, may reflect a reduction in programme duration but rarely account for further efficiencies, such as fewer subcontractors or less labour required on site. Other variables are more hidden and, as a result, frequently excluded from cost analyses altogether.

Examples of overlooked variables include:

- Weather impacts on productivity, which can be substantial in certain seasons or locations.
- Re-work, which represents one of the largest potential hidden costs. Although re-work incurs real expenditure, since it involves paying for the same work to be done twice, it is seldom recorded as a distinct cost by clients, main contractors, or subcontractors. Subcontractors often absorb the cost to preserve client relationships or to release retention payments, while clients may underestimate its impact beyond delays to effective property use. Nonetheless, subcontractors must ultimately recover these costs, embedding them within their pricing structures.

By failing to record or explicitly acknowledge these costs, stakeholders risk underestimating the true financial burden of traditional methods. MMC, by contrast, offers opportunities to reduce or eliminate some of these hidden costs, particularly re-work, through controlled production environments and repeatable processes. While factory-based methods may carry higher initial costs, the potential to reduce unrecorded or indirect costs suggests that overall project expenditure could be lowered if MMC is adopted and managed effectively.

Suggestions on How to Work in MMC

Throughout the book, the issues related to working with MMC have been raised, discussed, argued and addressed, as well as key suggestions in how to ensure, if you work in a project that uses MMC, you achieve maximum benefit and avoid the pitfalls. These have been summarised below.

Design Team and Contractor Selection

One of the key decisions in any project is to ensure the people who are involved in advising, designing, and building the project have the necessary strength in knowledge and experience to complete the project effectively. With a system and process that is relatively new to the industry this becomes far more important, as has been shown in several places through the book, lack of knowledge and skills in relation to a system, can lead to errors, rework and added expense. The appointment process in construction projects involves a series of analysis aspects followed by a very strict contractual agreement, typically using one of the off the shelf contracts.

The first part of the analyses is often a PQQ (Pre-Qualification Questionnaire) which explores much of the capabilities of the contractor, or consultant, including experience, staff knowledge, financial and insurance abilities. If the correct questions are asked at this stage, it will make the process of contractor selection later on much more successful. For this to work however, the client or the client's advisor also must know the sort of questions they need to ask.

It is also possible the decision to use MMC may come later in the design phase, such as Stage 1 or Stage 2 of the RIBA plan of work. This then makes the selection more complicated but equally more important as the later you leave the decision to use MMC into a project the harder it is to realise the benefits.

Ultimately, anyone who is reading this book, has increased their knowledge and abilities in MMC use, and as such the advice is for companies related to the design and build of projects to be more forth coming in their levels of abilities. Much in the same way that over the last few years it has become common place to advertise the level of BIM knowledge within the company, MMC knowledge should also be advertised. And client's knowledge and understanding needs to also be developed so that they can appreciate the ability to work in this sector.

Early Appointment of Contractors

Following on the from the suggestions above, this relates to the contractor, but also key subcontractors, and in particular the supplier, or off-site manufacturer. The key concept is that if you get contractors involved in the design stage of a project, they can have specific input in the buildability of a project, this is a key suggestion even on traditional builds. But due to the current lack of knowledge on MMC processes and products, it is likely to have more of an impact if it occurs on MMC projects.

As stated above, the earlier the decision to use MMC the easier the benefits can be realised, and if the right contractor along with the manufacturer is involved in the early aspects of the design, then the slight changes that can make a difference in aspects of the build can be incorporated into the overall project design. For example, Modular and Panelised off site systems work better with a standard size – if the standardised sizes are taken into account when doing the design then greater levels of accuracy, higher quality, and reduced cost can also be achieved. If the manufacturing plant has to adjust sizes, even slightly, to fit in with project designs, then this bespoke nature means resetting jigs in the factory, additional checks to ensure accuracy, and risk of human error in making these changes – it may also necessitate different delivery methods if the items become over size, which can add cost and logistical time constraints. This can lead to accidents in transit, or in storage. Therefore, designing a project to best meet the manufacturing needs has the potential for a large overall impact on product success. This is helped by early engagement with the contractors and their manufacturing supply chain.

Design Freeze

Another aspect that is key in an MMC project is the need to freeze the design earlier in the process than happens on a traditional project. The concept of a design freeze, and the benefits from doing it, would also be advantageous on a traditional project, but it is imperative on an MMC project due to the nature of the manufacturing process. The design freeze concept discussed several times in the book, is where a certain point in the design part of a project is met, and after that point no more changes can be made to the design. It is possible to make changes to the design following a design freeze, but all parties are aware that it will add significant cost and time to the project.

The use of BIM principles in the design will help in ensuring the design freeze can be achieved, as it allows for the client and other necessary stakeholders to immerse themselves in the virtual model of the project and ensure all decisions are made with the benefit of accurate understanding. As the MMC production takes a long time, and is typically started well in advance of the start on site time due to this, it is therefore important to plan this design freeze into the design programme from the start, or at least once the decision to use MMC has been made.

Finally, it is important that all stakeholders involved in the project are aware of the design freeze date – if one of the design team is working towards a different date, then information will not be ready and again errors can occur.

Financial Analysis of all Stakeholders

The impact on a traditional construction project of a supply chain member going into liquidation, or not being able to furnish the contract due to financial difficulties, can lead to delays and additional costs. In a traditional project, much of the supply chain is not unique and is readily available elsewhere – for example, if a timber merchant stops trading, there is often another timber merchant close by that can deliver the products. If a bricklayer goes into liquidation, again there is probably one close by that can pick up and finish the project – there may be a delay and additional cost, but it is relatively minor

If this happens in a supply chain that is in its infancy, as is the case with much of the off-site work, the ability to find an immediate replacement for supply and installation is very unlikely. If this happens then the impacts on time and cost can be very significant, even to the point of making the project no longer viable. Therefore, it is really important to carry out in-depth financial analysis of contractors, sub-contractors, and suppliers when working on an MMC project. This is does come at a cost, but it is key.

Increased Cost Control Process

Aligned with the above suggestion is the need for more strict cost control processes. This can start with a need for financial commitment from the manufacturer to ensure the project will complete, such as their depositing a sum of funds in a ring-fenced account, that will be used to produce the work –this type of financial security is used quite often with clients and contractors, depositing sums of money, set aside to ensure payment to the sub-contractor or supplier. When the supplier/sub-contractor does not have a secure working relationship with the client/

contractor, then this is a good financial safeguard. But there is nothing to stop it working the other way around – as a main contractor buying materials, you need certainty that they will turn up, one of the main reasons for materials not turning up is that the supplier has run out of funds. So, by ensuring they deposit funds sufficient to complete the order will give added security to the contractor. Note that it is not easy for the supplier to have all of this money in ready cash available to just deposit in a side account for the duration of the production cycle, so there may be a need to secure funds from a financier, or to have an insurance product to cover the cost. Either way it is a way of ensuring continued production for the client/contractor.

Another cost management mechanism is the use of warranties, guarantees, and performance bonds – these contractual mechanisms are also an insurance against low or poor performance on the part of suppliers and sub-contractors, as well as a consideration of the payment structure to supplier, sub-contractors and main contractors. Typically, there will be deposits paid to suppliers when they have a serious outlay far in advance – this means that clients will be paying out funds much earlier in the project than is typical. This will also leave the client exposed, or if the contractor is carrying the initial cost, it leaves the contractor exposed. Typical payment terms are that only work on site is paid for, and only once completed, and it is often on a monthly cycle on larger projects. It may be necessary to alter these standard practices for an MMC project to ensure that fewer of the stakeholders are unnecessarily carrying costs. Allowing in the contract for material off site to be paid for, ensuring quality checks and contractual adherence is carried out, even before the product has left the factory is another way to help.

The key point here is that standard financial control practices may have to change in order to ensure the industry is willing to move into the area of MMC, start to realise its benefits and develop the systems into a more standardised part of the sector.

Increased Quality Control Systems

One of the key benefits of using an off-site system or product is the perceived, or rather, possible, increase in quality. By utilising the increased control and controlled environment that is available in a fixed factory location, then the potential is for quality to increase. One of the key messages from Chapter 5 is that while quality increase is a possibility, it is not a given. For anyone wanting to work in this area it is important to understand what needs to be done to maximise the quality aspect. It is also important to recognise that the industry suffers from low quality output – it is recognised and accepted that latent defects happen and it is also recognised and accepted that human error creates mistakes on site that need to be fixed. However, this recognition and acceptance does not mean that it is right, and if a system can be used to increase the quality of output, especially if this can come at no additional cost, then it is incumbent on us to explore that as a viable option.

Chapter 5 explained some of the quality systems that can be used to increase quality in production and in installation and subsequent completion of the project. However, the key message is that the quality system should be in place, to ensure scrutiny is sufficient, and that this scrutiny occurs right from the start of the project. Such things as site teams observing factory production, quality checks and sign off at many more stages of a project than might happen traditionally, including design scrutiny and sign off.

While this level of detail has a cost implication, it will ensure that very costly re-work, and especially latent defect re-work, can be avoided. It is important to note that errors and re-work on MMC projects is likely to be more expensive, and so the mantra should be, if using MMC, do not skimp on quality processes. The benefits can be great, but errors will be costly.

Logistics and Installation Quality

Linked to the above, but of equal importance, is the concept of logistics quality – this is the transportation of materials to site and includes any temporary storage and then getting ready for installation (lifting by crane for example). This is a task that can lead to damage and errors even on traditional projects, but, as stated above, if these errors occur, then the replacement is typically reasonably easy to achieve. If damage occurs with MMC this is not always the case. Therefore, more care and consideration are needed when working with the logistics side of MMC. One key consideration is who should arrange the transportation, the storage, and the lifting into place.

With MMC the manufacturer is generally the specialist in the use of the product, and so it would be advisable to get this aspect organised by them as the chance of success will be greater, but also if it goes wrong there is more responsibility on the production team to put it right. There is a cost associated with this, however, and if the process is well understood and the risk is small, then it may be an idea for the on-site team to arrange the transport and installation directly.

When organising the delivery and installation, the same considerations will be needed that were discussed above – ensure the experience, quality, finances and insurances are all in place for each of the companies carrying out the work. There are, in this instance, many moving parts, which is again why it may be more beneficial to use the manufacturer themselves.

One of the key parts about installation is the coordination with the on-site team, and again the level of knowledge here is important. Factory production should be far more exact than that achievable on site, and yet these highly accurate products then have to be incorporated into on-site work. It is important that the on-site team, either main or sub-contractors, realise the high level of accuracy that is needed. If a slight error is made in the level of a foundation in a traditional project, this can be built out as the walls go up – in panelised systems or modular systems this cannot happen. Again, the key here seems to be knowledge, understanding, skills, and control checks.

Skills and Knowledge Development and Training

A running theme in all aspects of this book, and in fact one of the key reasons we wrote it, was the increase in skills and knowledge that is needed to ensure the benefits of MMC in projects is fully appreciated. The level of skills and knowledge can come from several directions, and it is important for those working in this sector or entering this sector to partake of as many as possible.

Reading

This text book, we feel at least, is a great way to start your journey into the use of MMC products. There is so much to consider that having a solid book in support and as a future reference guide is a great way to start or enhance knowledge. There are several text books out there, and more and more journal articles are being written on the subject. Reading and research to keep on top of developments and ideas relating to the sector is key.

CPD

Continuing professional development is a way of ensuring you can keep on top of these developments, in an efficient way. It usually requires some form of professional networking event, trade show, industry show or presentation from industry experts. There may be a cost involved in attendance, either from ticket entry prices or just the lost time to attend. But it can short circuit some of the key aspects that need to be considered when moving into, or working within, the sector using MMC products.

Education

Selecting a training facility for you or your employee's development is a really important way to start, as long as they are at that stage of development, of course. Choosing technical education such as trade qualifications that provide an insight into how to work with MMC are not common, although there is a move within the Further Education (FE) sector to provide more insight into this growing sector. This will definitely help with the concept of linking and integrating the on-site skills with the delivered products as discussed above.

Another key aspect is the levels of management, control, and integration needed within the professional sector, so again it is important to explore Higher Education (HE) institutions that will provide this level of knowledge and insight so that those in decision making positions in the future have the understanding that is needed. It is becoming more common to have degree courses that have links to MMC included, but having a specific MMC course is a way off yet. However, an exploration of modules on a course and seeking ones that cover aspects of MMC is a good place to start, until the sector grows and becomes more mainstream.

This is the same for apprenticeships – at all levels, including degree apprenticeships, at the moment there are no standards that specifically deal with MMC or off-site construction methods, but when seeking look at universities and Further Education institutions that have the pedigree and background to ensure those leaving with the qualification are best prepared to work within this growing sector.

Another key consideration, and one that the industry has historically struggled with is the concept of "we change what is successful" or "we do it this way, as we have always done it this way". The fact that decision makers in industry are typically people who were educated many years ago, with the technologies and processes that were well established then, means that despite the fact that we are in a technologically advancing industry, there is a reluctance in industry to make changes. Therefore, it is key that those being trained in up-to-date methods are listened to and used to ensure that the industry is more likely to develop quicker than it has in the past.

Experience

A final way to evolve knowledge is through experience – this can be your own experience or by listening to and learning from the experience of others. The drawback to the first way is that it can take a lot of time and quite a few mistakes in order to learn the right way to do something. The drawback to the second way is that the experience quite often has a personal perception attached, which may skew the facts. However, experience is beneficial in that it is lived experience and, due to its real-life nature, it creates a solid link to what does and does not work, whereas the other examples given above have a step removed between the experience and the person gaining the knowledge.

Public Perception Development

One of the key aspects to ensuring MMC is used more in future works is the necessary adjusting of public perception. The term "prefab" still has negative connotations in the UK, and wider, as it is perceived the speed associated with prefabricated builds leads to lower quality and care. The links to trailer parks, as seen in America and in some parts of the rest of the world (in the UK they are called residential parks) also carry with them a feeling of reduced quality, lack of longevity, and a cheap alternative to a permanent brick-built, on-site property.

But as has been discussed throughout this book, the MMC processes are not the same as these historic building processes. Modern MMC can have much higher quality than that seen on site, they can produce buildings with much greater energy efficiency than those built on site. Yes, they are quicker, but this should not be seen as a lack of quality, and this mis-perception of the public needs to change. Over time it will – in the same way that trussed roofs are now the accepted norm on house builds whereas, 40 years ago, the public were very sceptical. But is there something we can do to speed up this process of acceptance?

One of the kay ways to do this is with greater investment from government, not necessarily financial investment, but emotional investment in the process. If the government were to look at public funded projects being carried out using MMC techniques and extoll the benefits, this would help the public see for themselves, and become more accepting more quickly.

One of the public issues, in the UK and in many countries across the world, is the housing crisis, the fact that we are not building enough homes. However, unless the government takes a lead in solving this, it will not be resolved by housing developers – if they did, they would lose profit, and this is not in their own interest to do so. Therefore, if the government invested in social housing, using MMC methods, it would show that high-quality homes can be built quickly and efficiently and this, hopefully, would help with public perception.

The housing sector, in particular, is plagued by bad press relating to poor new home quality standards, showing the issue relating to on-site building processes. If investment in MMC could show that there are no latent defects in properties built this way, then the public perception, if marketed correctly would also change.

The key issue here is that clients/developers commission or speculatively build properties that are desired by the end user. If the end user is sceptical about a product, the client/developer will not build it. This is one of the reasons you will see MMC being used extensively in the leisure industry (hotels, etc.) where the end user has no idea how the hotel was built, but the hotel chain appreciates the speed and quality afforded by MMC. The housing industry is slower to realise these benefits, because the public buying the homes have often seen the speed with which they have been built and are sceptical of the associated quality.

Develop Cost Models to Encapsulate all Cost Variables

One of the key reasons for writing this book was to explore the concept of cost, and while there is no easy or direct solution to this, due to the unique nature of our industry, the book has advocated the development of a cost model that allows for the inclusion of the necessary elements and variables in a construction project. While there are several models out there, and even some software companies claiming to do this, the true nature of off-site construction and its impact on individual projects has yet to be completed.

Until the ultimate piece of software is developed that takes into account all of the cost variables and allows industry to ascertain the cost of these variables in relation to their own unique situation, and in their own unique project, then, for now, it is important that those making the decisions take into account the variables and make their "best guess" as to its impact.

Until this allowance is made, the industry is unlikely to move forward and accept the benefits that can come from embracing MMC. It has been shown in several of the chapters in the book, but in particular in Chapter 7, that cost is the key factor in decision making. It appears that clients are unlikely to worry about quality and time if the project is too costly. There is always a balance to be struck, but once the cost has gone beyond the realms of affordability, then the time or quality will be adjusted accordingly, or the project will not go ahead.

It was also discussed that due to the overheads and set up costs associated with the factories needed for MMC manufacturing, the production and installation costs are generally considered prohibitive when compared to traditional on-site costs. The cost of re-work (which can be extensive), the impact of weather on productivity, the cost of health and safety the reduction in preliminary requirements, etc., along with many other of the 70+ cost variables discussed in Chapter 7 – all show that the question of cost is not as simple as is often suggested. While the calculation of these costs requires a lot more investigation and research from industry, it is suggested that if MMC systems are used correctly and effectively, as intended and as suggested in this section of the book, then the reduction in cost should make it competitive, if not advantageous, as a production method.

This would allow clients to benefit from all of the other advantages from MMC – increased quality, reduced production time, greater cost certainty, etc. – with no or even reduced cost. But a word of warning to end with: this is with the proviso that the industry starts to use manufacturing processes in the way they were intended, and not simply taking the mistakes of construction and moving them indoors. MMC processes have to be used properly and, to gain the full benefit, this takes time, knowledge, and perseverance, and some investment. But, if done correctly, the benefits to clients, and the industry, should be realised.

Summing All of This Up

As this book draws to a close, it is worth reflecting on the broader significance of Modern Methods of Construction. Beyond procurement strategies, cost considerations, and programme efficiencies, MMC represents a cultural and industrial shift in how we think about building. It challenges traditional assumptions, not by discarding them, but by reframing them within a system that prizes predictability, safety, sustainability, and long-term value.

For decision-makers, the central task ahead is not simply whether to adopt MMC, but how to embed its principles in a way that is commercially viable, socially responsible, and technically robust. That means engaging supply chains early, setting clear client requirements that reflect the realities of factory-based manufacture, and investing in the skills needed to design for manufacture and assembly. It also means acknowledging the risks and uncertainties that remain, but managing them with evidence, data, and collaboration, rather than allowing them to block progress.

The opportunity is clear. Projects delivered through MMC can achieve gains in cost certainty, programme reliability, quality assurance, and energy performance. Communities can benefit from reduced disruption, quicker access to homes and facilities, and lower carbon impacts. The construction workforce can benefit from safer working environments and more stable employment. But realising these gains requires maturity: in procurement, in planning, in client expectations, and in professional practice.

Over the coming years, further research will refine our understanding of MMC's true cost dynamics and its social and environmental impact. Until then, it falls to practitioners, policymakers, and clients to make informed decisions that reflect both the challenges and the benefits. If this book has one final message, it is this: Modern Methods of Construction are not an alternative to "real building", they are part of the future of building itself. The sooner we learn to use them wisely, the greater the rewards will be.

Index

Note: Page numbers in **bold** refer to tables.

For Product Safety Concerns and Information please contact our EU representative GPSR@taylorandfrancis.com
Taylor & Francis Verlag GmbH, Kaufingerstraße 24, 80331 München, Germany

www.ingramcontent.com/pod-product-compliance
Ingram Content Group UK Ltd.
Pitfield, Milton Keynes, MK11 3LW, UK
UKHW050050200926
475702UK00019B/2763
* 9 7 8 1 0 3 2 9 1 6 0 6 4 *